JOSEPH S. CHIARAMONTE, M.D.
649 MONTAUK HWY.
BAY SHORE, N.Y. 11706

J Chiaramonte, M.D.
7/19/74

11/16/02 —

I give this book to
Janet Wilshe ~~her~~ a
great clinical professional
Good Luck
Dr. C.

JOSEPH S. CHIARAMONTE, M.D.
949 MONTAUK HWY.
BAY SHORE, N.Y. 11706

Handbook of Differential Diagnosis

Volume 1

THE CHEST

A PUBLICATION OF
ROCOM™ PRESS

ROCOM DIVISION
Hoffmann-La Roche Inc., Nutley, New Jersey

Copyright © 1967, 1968, 1974 Hoffmann-La Roche Inc. All rights reserved.

PREFACE

The HANDBOOK OF DIFFERENTIAL DIAGNOSIS was developed to provide valuable assistance to both practicing physician and medical student in arriving at an accurate diagnosis from the starting point of the patient's presenting signs or symptoms.

Each volume concentrates attention on a major body region or system, and discusses the diagnostic possibilities to be considered when one is confronted by a specific abnormality within that region or system. The manner in which the signs or symptoms present in each disorder is outlined in telegraphic fashion, along with relevant associated findings that may contribute useful information toward the differential diagnosis. Original medical art in full color, as well as x-rays, scans and tracings from the collections of major teaching institutions supplement the text. For rapid review, at the end of each section is a table summarizing the signs, symptoms, associated findings, key laboratory data, etc., of those conditions or disease states presented in the section.

Originally published by Roche Laboratories as a series of individual booklets, the HANDBOOK covers disease entities physicians are most likely to encounter in daily practice, as well as those which present a particular challenge in differentiation. Developed after extensive research and up-dated where necessary for this edition, the HANDBOOK incorporates the views of many recognized authorities and reflects today's accepted medical practice.

The content, the manner in which it is presented and the accompanying illustrations all combine to provide a concise, convenient and reliable source of reference or review for the physician's library.

CONTENTS

Preface 3

DIFFICULTY IN BREATHING

Introduction 9
Laryngeal Obstruction 10
Pulmonary Emphysema 11
Pneumonia 12
Pulmonary Fibrosis 13
Pulmonary Infiltrative Lesions ... 14
Bronchial Asthma 15
Pulmonary Edema of Left Ventricular Failure 16
Pulmonary Edema of Mitral Stenosis 17
Obesity 18
Acidotic Hyperpnea 19
Hyperventilation Syndrome 20
Pneumothorax 21
Table of Differential Diagnosis 22-25

MIDLINE CHEST PAIN

Introduction 27
Angina Pectoris, Coronary Insufficiency, Myocardial Infarction 28, 29
Pulmonary Hypertension 30
Acute Pericarditis 31
Aortic Dissection 32
Pulmonary Embolus 33
Pneumomediastinum 34
Esophagitis 35
Esophageal Spasm 36

Anxiety 37
Cervical Outflow Syndromes 38
Musculoskeletal Disorders 39
Table of Differential Diagnosis 40-43

CHEST PAIN ON BREATHING

Introduction 45
Acute Pleurisy 46
Pneumonia 47
Lung Abscess 48
Pulmonary Infarction 49
Pneumothorax 50
Cancer 51
Pleurodynia 52
Trichinosis 53
Herpes Zoster 54
Diaphragmatic Flutter 55
Actinomycosis 56
Cholecystitis 57
Table of Differential Diagnosis 58-61

COUGH

Introduction 63
Chronic Bronchitis 64
Thoracic Aortic Aneurysm 65
Bronchogenic Carcinoma 66
Pulmonary Tuberculosis 67
Bronchiectasis 68
Pneumococcal Lobar Pneumonia 69
Klebsiella Pneumonia 70

Continued on next page.

COUGH (continued)

Micoplasma Pneumonia 71
Lung Abscess 72
Acute Tracheobronchitis 73
Pulmonary Infarction 74
Acute Pulmonary Edema 75
Table of Differential Diagnosis 76-79

PULMONARY RALES

Introduction 81
Description of Pulmonary Rales —
Dry, Moist 82
Acute Pulmonary Edema 83
Pulmonary Infarction 84
Lobar Pneumonia 85
Bronchiectasis 86
Pulmonary Tuberculosis 87
Acute Tracheobronchitis 88
Chronic Bronchitis 89
Bronchial Asthma 90
Lung Abscess 91
Bronchopneumonia 92
Mycoplasma Pneumonia 93
Table of Differential Diagnosis 94-97

PLEURAL EFFUSION

Introduction 99
General Characteristics 100, 101
Congestive Heart Failure 102
Pulmonary Infarction 103

Bacterial Pneumonia 104
Tuberculosis 105
Malignancy 106
Lymphoma 107
Cirrhosis of Liver, Meig's Syndrome,
Nephrotic Syndrome 108, 109
Systemic Lupus Erythematosus 110
Subdiaphragmatic or Hepatic Abscess . . . 111
Table of Differential Diagnosis 112-115

GALLOP RHYTHM

Introduction 117
General Principles 118, 119
Arteriosclerotic Heart Disease 120
Hypertensive Heart Disease 121
Aortic Stenosis 122
Aortic Insufficiency 123
Mitral Insufficiency 124
Pulmonary Hypertension 125
Primary Myocardial Disease 126
Idiopathic Hypertrophic Subaortic
Stenosis 127
Pulmonic Stenosis 128
Constrictive Pericarditis 129
Table of Differential Diagnosis 130-133

SYSTOLIC MURMUR

Introduction 135
Mitral Insufficiency 136
Tricuspid Insufficiency 137
Valvular Aortic Stenosis 138

Idiopathic Hypertrophic Subaortic Stenosis 139	
Supravalvular Aortic Stenosis 140	
Pulmonic Stenosis 141	
Tetralogy of Fallot 142	
Coarctation of the Aorta 143	
Interventricular Septal Defect 144	
Interatrial Septal Defect 145	
Disease of Papillary Muscles or Chordae Tendinae 146	
Murmurs in the Normal Subject 147	
Table of Differential Diagnosis 148-151	

DIASTOLIC MURMUR

Introduction 153
Mitral Stenosis 154
Left Atrial Myxoma 155
Mitral Insufficiency 156
Aortic Insufficiency of Rheumatic Heart Disease 157
Aortic Insufficiency of Syphilitic Heart Disease 158
Aortic Insufficiency of Rheumatoid Arthritis 159
Aortic Insufficiency of Marfan's Disease 160
Tricuspid Stenosis 161
Pulmonic Insufficiency 162
Patent Ductus Arteriosus 163
Extracardiac A-V Shunting (other than Patent Ductus Arteriosus) 164
Septal Defects 165
Table of Differential Diagnosis 166-169

ABNORMAL HEART SOUNDS

Introduction 171
Generation of Heart Sounds 172, 173
Loud S_1 due to
   Increased Contractility 174
   Mitral Stenosis 174
Soft S_1 due to
   Decreased Contractility 175
Varying intensity of S_1 due to
   Atrial Fibrillation 175
   Complete Heart Block 176
   Ventricular Tachycardia 176
Loud A_2 due to
   Systemic Hypertension 177
Loud P_2 due to
   Pulmonary Hypertension 177
Split S_2 due to
   Interatrial Septal Defect 178
   Pulmonic Stenosis 178
   Left Bundle Branch Block 179
   Right Bundle Branch Block 179
   Aortic Stenosis 180
   Idiopathic Hypertrophic Subaortic Stenosis 180
   Severe Heart Failure 181
   Acute Myocardial Infarction 181
   Pulmonary Fibrosis 182
   Pulmonary Emphysema 182
   Patent Ductus Arteriosus 183
   Mitral Insufficiency 183
Table of Differential Diagnosis 184-187

DIFFICULTY IN BREATHING

Difficulty or discomfort in breathing, when described by the patient, is the symptom of dyspnea; when apparent to the examiner, it is the physical sign of labored breathing. The symptom and the sign usually occur simultaneously, but either may be present without the other. Both are caused by a variety of diseases of the larynx, bronchi, lungs, heart, nervous system and of acid-base balance.

A clear understanding of the situation in which the difficult breathing occurs, the character of breathing and the associated signs and symptoms in each causal condition will facilitate correct diagnosis.

The diagnostically important symptoms and signs of twelve major causes of difficult breathing are described individually and then compared in tabular form.

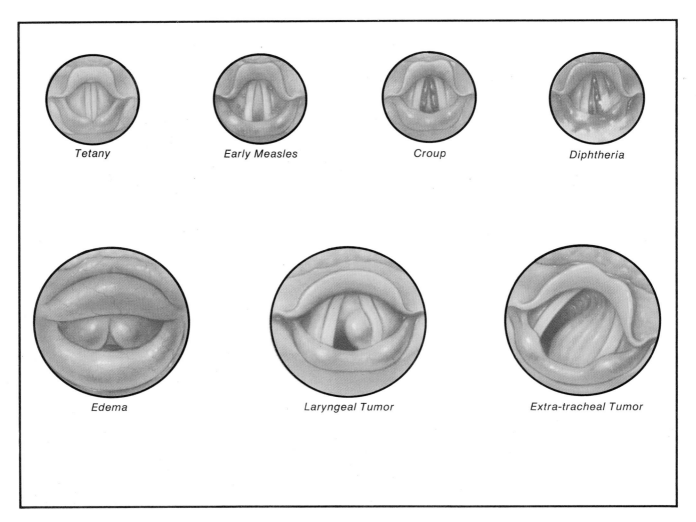

Tetany · Early Measles · Croup · Diphtheria
Edema · Laryngeal Tumor · Extra-tracheal Tumor

Difficulty in breathing due to LARYNGEAL OBSTRUCTION

Laryngeal obstruction, an important cause of breathing difficulty, may be the result of various conditions. Partial obstruction of the larynx is often due to edema, which may be caused by allergic reactions, by physical trauma or by chemical insult. Spasm during anesthesia or tetany may also cause laryngeal obstruction. Other possible causes are the presence of tumors or of foreign bodies accidentally lodged in the throat. The laryngitis of croup, measles or diphtheria may also create obstruction.

- **Dyspnea** persistent.
- **Present only on inspiration**, except in terminal suffocation.
- **Position** has no effect on dyspnea; no orthopnea.
- **No paroxysmal nocturnal dyspnea.**
- **Inspiration** usually prolonged with harsh high-pitched noise.
- **Stridor** (crowing sound) may be audible at a distance.
- Breathing becomes rapid in terminal suffocation, usually with expiratory distress.
- Only long stridulous inspiration sounds; no rales or change in percussion note.

ASSOCIATED FINDINGS

- Barking cough.
- Hoarseness and even aphonia.
- Typically, indrawing of suprasternal notch, supraclavicular and upper intercostal spaces on inspiration — due to marked increase of negative pressure in chest.
- Cyanosis may be marked.
- When obstruction is due to allergic edema, urticaria and angioneurotic edema are often present.
- When obstruction is due to laryngitis of measles, Koplik's spots are usually seen.
- Stridulous breathing in sarcoidosis can be due to laryngeal granuloma rather than pulmonary dyspnea.

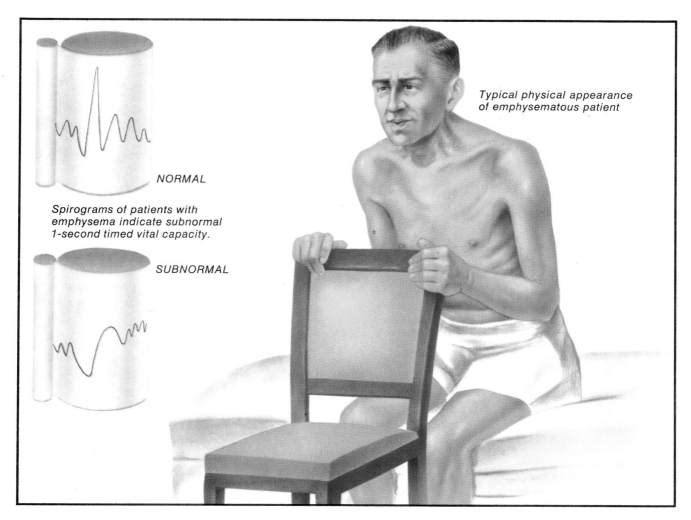

Typical physical appearance of emphysematous patient

Spirograms of patients with emphysema indicate subnormal 1-second timed vital capacity.

NORMAL

SUBNORMAL

Difficulty in breathing due to
PULMONARY EMPHYSEMA

Pulmonary emphysema, clinically defined as chronic lower airway obstruction, is most often seen in middle-aged males with a history of asthma or bronchitis. It may be secondary to fibrosing pulmonary diseases such as tuberculosis or silicosis. The transition from simple chronic bronchitis to emphysema is gradual and difficult to define.

- Dyspnea only on exertion in mild cases.
- Dyspnea also at rest in advanced cases and when bronchitis is acute.
- Orthopnea and paroxysmal nocturnal dyspnea usually absent; patient may awaken coughing due to pooled secretions.
- Patient often leans forward and purses lips to help breathing.
- Breathing labored but not always rapid; expiration very prolonged.
- Breath sounds, when audible, are faint and harsh without vesicular quality.
- Expiratory rhonchi (coarse rales) and wheezes (musical rales) usually present; ordinarily no fine rales.

ASSOCIATED FINDINGS

- Cough always present, usually nonproductive; purulent sputum on awakening or when bronchitis is acute.
- Cyanosis and drowsiness from CO_2 retention in advanced cases or with superimposed infection.
- Fixed chest wall; barrel chest develops in the stocky, chicken breast in the slim.
- Marked wasting from loss of appetite.
- Hyperresonant lungs; vocal fremitus, breath sounds and spoken voice impaired or absent.
- Polycythemia, pulmonary hypertension and cor pulmonale may develop — with persistent cyanosis.
- Chest x-ray typically reveals hyperlucency and depressed diaphragm; may be normal even with significant emphysema.
- Maximum breathing capacity more abnormal than vital capacity.

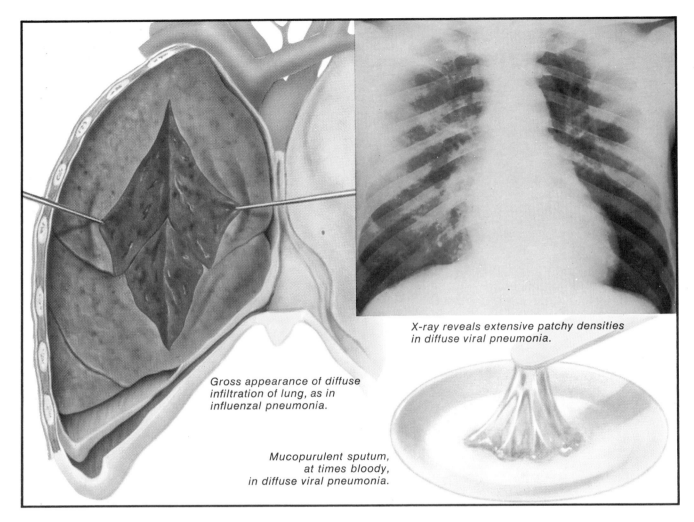

Gross appearance of diffuse infiltration of lung, as in influenzal pneumonia.

X-ray reveals extensive patchy densities in diffuse viral pneumonia.

Mucopurulent sputum, at times bloody, in diffuse viral pneumonia.

Difficulty in breathing due to PNEUMONIA

Marked dyspnea may be produced by several types of pneumonia. It is prominent, for example, in the diffuse viral pneumonias that complicate epidemic influenza and in varicella pneumonia in the adult. Dyspnea is also present in bacterial lobar pneumonia, but is not usually a feature of bacterial bronchopneumonia or of the patchy pneumonias due to virus or mycoplasma (primary atypical pneumonia). When it is present in these pneumonias, the cause is usually underlying heart or lung disease.

- Severe dyspnea at rest in diffuse viral pneumonia; resting dyspnea also common in bacterial lobar pneumonia.
- No orthopnea or paroxysmal nocturnal dyspnea.
- Breathing rapid and shallow; in bacterial lobar form breathing is often jerky from pleural pain, usually accompanied by expiratory grunting.
- Prolonged expiration and marked wheezing in influenza pneumonia.
- Diminished breath sounds, diffuse rales and wheezes in diffuse viral pneumonias; fine and medium inspiratory rales in bacterial lobar form.

ASSOCIATED FINDINGS

- Fever, prostration, chest pain prominent in bacterial lobar pneumonia.
- Varicella pneumonia begins 2 to 5 days after rash.
- Cyanosis common in diffuse viral pneumonias; variable in bacterial lobar forms, often marked in staphylococcal lobar pneumonia complicating influenza.
- Severe cough.
- Sputum purulent and often bloody in diffuse viral forms; mucopurulent in bacterial lobar forms; "rusty" in pneumococcal pneumonia.
- Signs of consolidation over involved lobe in bacterial lobar pneumonia — dullness to percussion, increased tactile fremitus, bronchial breathing; not present in diffuse viral forms.
- Dullness to percussion from pleural effusion in varicella pneumonia.
- X-rays show diffuse bilateral density in diffuse viral pneumonia; opacification of consolidated lobe in bacterial lobar pneumonia.

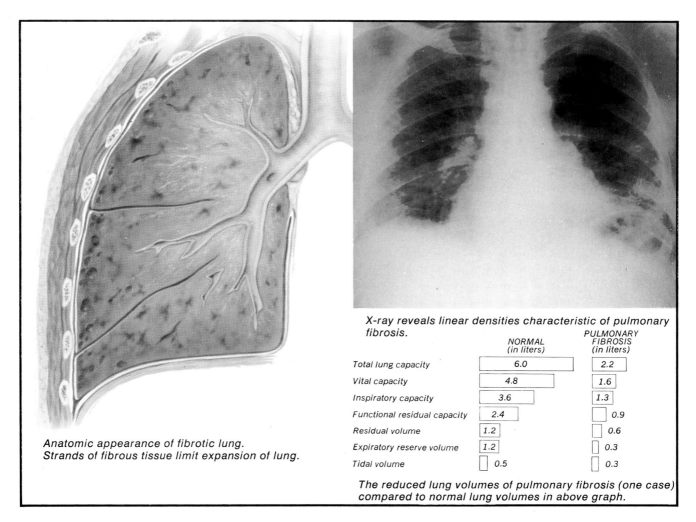

Anatomic appearance of fibrotic lung. Strands of fibrous tissue limit expansion of lung.

X-ray reveals linear densities characteristic of pulmonary fibrosis.

	NORMAL (in liters)	PULMONARY FIBROSIS (in liters)
Total lung capacity	6.0	2.2
Vital capacity	4.8	1.6
Inspiratory capacity	3.6	1.3
Functional residual capacity	2.4	0.9
Residual volume	1.2	0.6
Expiratory reserve volume	1.2	0.3
Tidal volume	0.5	0.3

The reduced lung volumes of pulmonary fibrosis (one case) compared to normal lung volumes in above graph.

Difficulty in breathing due to PULMONARY FIBROSIS

Pulmonary fibrosis may be caused by pneumoconiosis from industrial dust inhalation, sarcoidosis or tuberculosis of the lung with residual scarring, and multiple small pulmonary infarctions. A progressive form of unknown etiology—Hamman-Rich syndrome—is fatal in a year or two. Fibrosis sometimes follows pneumonia and radiation, but this type of condition is usually not extensive enough to produce the dyspnea that occurs with other forms of pulmonary fibrosis.

- Dyspnea initially only on exercise, later persistent; no orthopnea or paroxysmal nocturnal dyspnea.

- Rapid, shallow breathing even before patient is aware of dyspnea.

- Breathing loud and bronchial unless pleurae are fibrotic.

- Prolonged expiration, wheezing and barrel chest in silicosis — because of associated emphysema.

ASSOCIATED FINDINGS

- Usually nonproductive cough.
- Usually cyanotic only on exercise.
- Chest expansion diminished; diaphragm moves little.
- Inspiratory retraction of rib interspaces when fibrosis is marked.
- Fine crepitant basal rales; rhonchi and wheezes rare.
- Loud P_2, xiphoid presystolic gallop (pulmonary hypertension) and right heart failure possible without persistent cyanosis.
- Cyanosis and pulmonary hypertension are persistent when they occur.
- X-ray reveals diffuse linear densities, except in silicosis where fibrosis conglomerates.
- "Egg-shell" calcification of hilar nodes in silicosis.
- Vital capacity markedly reduced; maximum breathing capacity less so.
- Typical asbestos bodies in sputum with asbestosis.

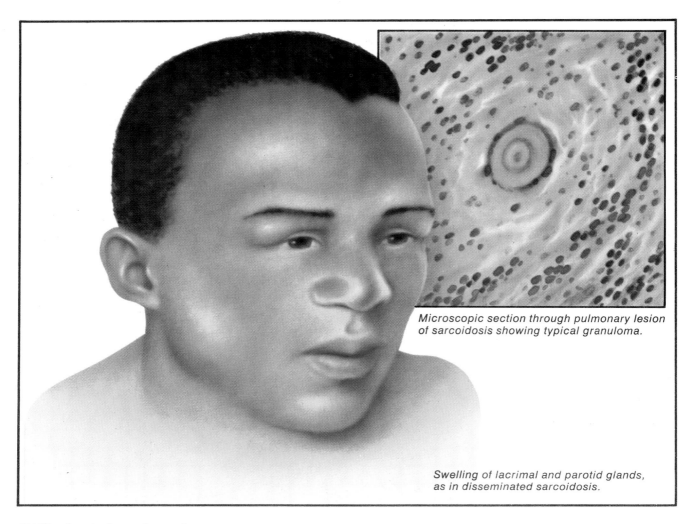

Microscopic section through pulmonary lesion of sarcoidosis showing typical granuloma.

Swelling of lacrimal and parotid glands, as in disseminated sarcoidosis.

Difficulty in breathing due to
PULMONARY INFILTRATIVE LESIONS

Respiratory distress may be caused by a variety of pulmonary infiltrative lesions. Granulomas due to miliary tuberculosis, sarcoidosis, berylliosis or fungal infections may fill the interstitial lung spaces and cause difficulty in breathing. Pulmonary lymphatic cancers, Hodgkin's disease and other lymphomas, and connective tissue disorders such as scleroderma may also cause respiratory distress. Similar symptoms and signs are occasionally produced by intra-alveolar deposits (pulmonary alveolar proteinosis).

- Dyspnea persistent; unrelated to position.
- Hyperventilation and PCO_2 reduction are common.
- Rapid, shallow breathing, without audible wheezing.
- Harsh breath sounds and few fine rales may be the only auscultatory findings.
- Prolonged expiration and musical rales *not* present.

ASSOCIATED FINDINGS

- Usually nonproductive cough.
- Cyanosis — first on exercise, later even at rest; may not be fully corrected by oxygen.
- Pulmonary sarcoidosis and miliary tuberculosis may show negative tuberculin tests, fever, little cough and granuloma in liver.
- Sarcoid often causes lacrimal and parotid gland swelling, uveitis and granuloma in skeletal muscle.
- Pleural and pericardial effusions, meningitis and peritonitis often seen with miliary tuberculosis.
- Fungal infection may result in diffuse pulmonary infiltration.
- X-ray reveals patchy densities throughout lungs, "softer" than in pulmonary fibrosis; sarcoid shows hilar nodes on x-ray.
- Carcinoma spreading through pulmonary lymphatics can present same clinical and x-ray signs; primary lesion often in stomach.
- Hodgkin's disease and other lymphomas can cause similar pulmonary findings; nodes or tumors present elsewhere.

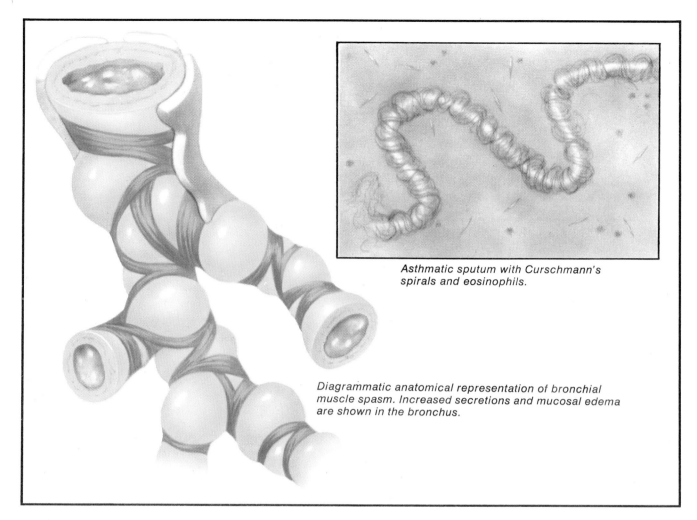

Asthmatic sputum with Curschmann's spirals and eosinophils.

Diagrammatic anatomical representation of bronchial muscle spasm. Increased secretions and mucosal edema are shown in the bronchus.

Difficulty in breathing due to BRONCHIAL ASTHMA

An important cause of respiratory distress is bronchial asthma. In children and young adults this condition is usually allergic in origin, and in many cases a familial history may be found. If onset is after age 40, there is often a history of chronic bronchitis, but usually no allergic or familial component.

- Recurrent acute attacks of dyspnea and wheezing, often accompanied by tightness and burning in chest.
- Sometimes paroxysmal attacks when patient is sleeping.
- Breathing readily heard; inspiration is short, expiration greatly prolonged and laborious.
- In brief attacks patient sits up and frequently leans over table or chair back.
- In severe attacks breathing becomes rapid despite prolonged expiration.
- In prolonged attacks (status asthmaticus) patient lies prostrate, uses accessory muscles less and becomes cyanotic.
- Expiratory breath sounds in asthma longer and much higher pitched than in bronchial breathing.
- The higher pitched the wheezing, the more severe the attack and the obstruction.
- Usually coarse musical rales (wheezes); fine rales occasionally present.

ASSOCIATED FINDINGS

- Attacks often preceded by sneezing or itching of skin of anterior chest.
- Dyspnea followed by jerky nonproductive cough.
- Patient improves when cough becomes productive and tenacious sputum is raised.
- Cyanosis may be marked.
- Asymptomatic between attacks.
- Chest hyperresonant and hyperinflated; lung bases low.
- Sputum tenacious and mucoid; many eosinophils.
- X-ray normal or shows hyperlucent lung fields.
- Sudden occurrence of left-sided cardiac failure may closely simulate asthma; labored breathing may prevent auscultation of heart.

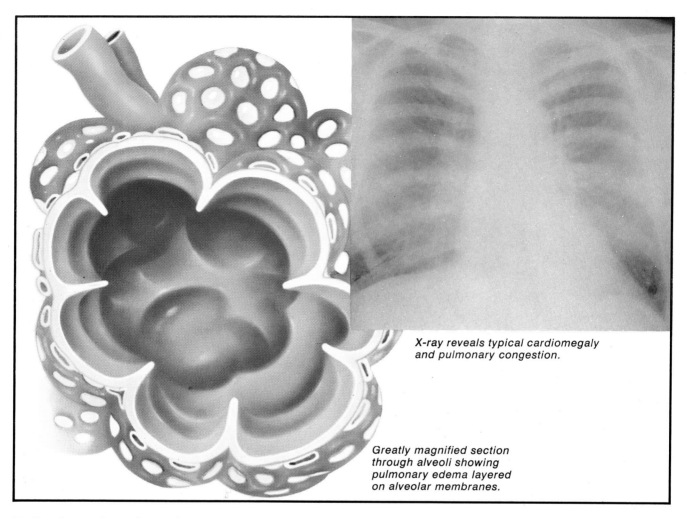

X-ray reveals typical cardiomegaly and pulmonary congestion.

Greatly magnified section through alveoli showing pulmonary edema layered on alveolar membranes.

Difficulty in breathing due to PULMONARY EDEMA OF LEFT VENTRICULAR FAILURE

The shortness of breath of heart failure, paroxysmal nocturnal dyspnea, acute pulmonary edema and cardiac asthma reflect the increased work of inflating stiffened, congested lungs associated with left ventricular failure.

- Early, dyspnea only on exertion.

- Later, paroxysmal nocturnal dyspnea causes patient to awaken after 1 or 2 hours' sleep; must sit or stand for 15 to 20 minutes to improve breathing.

- Still later, dyspnea as soon as patient lies flat, or nearly so — orthopnea.

- Dyspnea present regardless of activity or position in persistent or severe pulmonary edema.

- Bilateral inspiratory crepitant rales.

- Breathing rapid, shallow and labored.

ASSOCIATED FINDINGS

- Nonproductive cough in all degrees of pulmonary edema; preterminally, frothy pink sputum.

- Peripheral cyanosis can result from low cardiac output; marked arterial unsaturation uncommon.

- Profuse sweating can occur as in bronchial asthma.

- In chronic heart failure, rales may be absent with other signs of pulmonary edema present.

- Tachycardia and protodiastolic gallop at cardiac apex.

- With sinus rhythm presystolic gallop often heard.

- Distended neck veins, enlarged liver and dependent edema of right heart failure need not be present.

- Pleural effusion usually occurs only when both right and left ventricular failure are present.

- X-ray reveals cardiomegaly and pulmonary vascular congestion.

- Acute episodes of expiratory wheezing without fine rales can occur in pulmonary edema, simulating bronchial asthma — especially in patients with asthmatic history; apical gallop rhythm, if not obscured by wheezing, will clarify diagnosis of cardiac asthma.

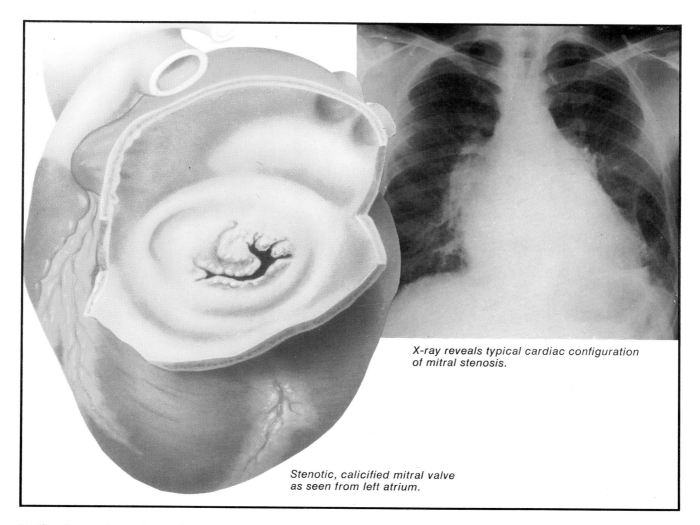

X-ray reveals typical cardiac configuration of mitral stenosis.

Stenotic, calicified mitral valve as seen from left atrium.

Difficulty in breathing due to PULMONARY EDEMA OF MITRAL STENOSIS

Shortness of breath is typical of the pulmonary edema accompanying mitral stenosis, a condition seen not only in young adults (particularly women) but also in the aged. Less than half of the patients have a history of rheumatic fever. High cardiac output in pregnancy, anemia and fever may intensify the symptoms.

- Exertional dyspnea first symptom; paroxysmal dyspnea may be triggered by emotional stress and onset of atrial fibrillation.
- True paroxysmal nocturnal dyspnea uncommon; but patient may become dyspneic when he rolls off propped-up pillows (orthopnea).
- Dyspnea may not be prominent when pulmonary hypertension is marked and cardiac output small, although patient is cyanotic.
- Rapid, shallow breathing.

ASSOCIATED FINDINGS

- Usually nonproductive cough; sometimes clear sputum with many "heart failure" cells.
- Frothy pink sputum a late finding; gross hemoptysis may occur.
- Peripheral cyanosis; true cyanosis in severe cases.
- Dysphagia and hoarseness, when present, due to enlarged atrium.
- Superimposed bronchitis common; wheezing may simulate bronchial asthma.
- Bilateral inspiratory crepitant rales.
- Pleural effusion may be present — dullness to percussion, diminished breath sounds and tactile fremitus at lung bases.
- Typically, tachycardia and apical diastolic rumble; often opening snap of mitral valve in early diastole.
- Irregular heart rate of atrial fibrillation common.
- X-ray shows pulmonary vascular congestion, enlarged pulmonary arteries, double density of enlarged atrium.
- In initial stages PA chest x-ray normal; oblique view needed to reveal left atrial enlargement.

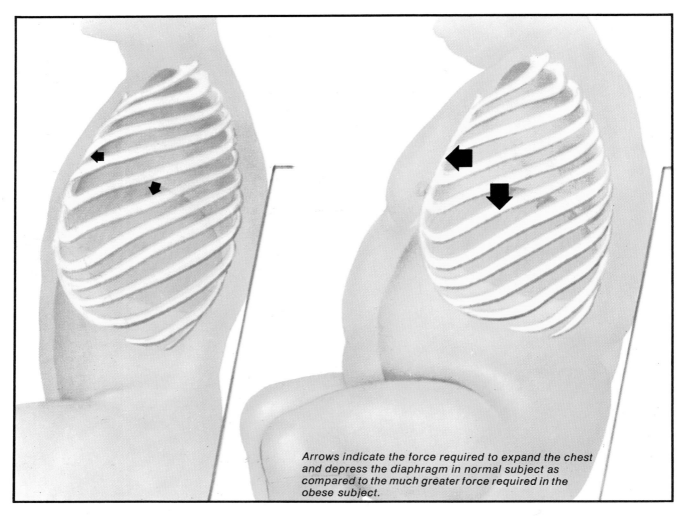

Arrows indicate the force required to expand the chest and depress the diaphragm in normal subject as compared to the much greater force required in the obese subject.

Difficulty in breathing due to OBESITY

Obesity can cause breathing difficulties because of the increased work required to raise the heavy chest and displace the large abdomen during inspiration. The same mechanism may cause dyspnea during pregnancy or with ascites. The dyspnea resulting from myasthenia, quadriplegia, muscular dystrophy and polio is physiologically similar but results from muscular weakness.

- Dyspnea on exertion.
- Dyspnea also in supine position if obesity is extreme.
- No paroxysmal nocturnal dyspnea.
- Breathing rapid and shallow.
- Breathing may become irregular when PCO_2 is high.

ASSOCIATED FINDINGS

- Physical examination reveals chest entirely normal.
- No serious ventilatory insufficiency unless lung disease is also present.
- If patient is bronchitic, even moderately, breathing is so difficult that gas exchange becomes inadequate for normal blood gas concentrations — alveolar hypoventilation.
- Obese hypoventilators are often cyanotic and polycythemic because of low PO_2, and drowsy because of high PCO_2.
- Loud P_2 and xiphoid presystolic gallop may occur if pulmonary hypertension is present.
- With right heart failure, distended neck veins, hepatomegaly and edema may occur.
- Cyanosis, pulmonary hypertension and heart failure are all reversible with weight loss.
- Hematocrit and serum bicarbonate often elevated.
- X-ray normal.
- Arterial blood gas analyses and pulmonary function tests necessary to establish diagnosis.

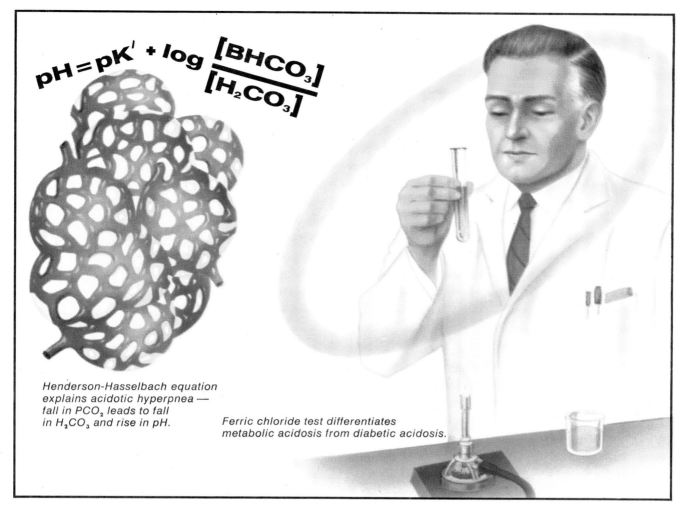

Henderson-Hasselbach equation explains acidotic hyperpnea — fall in PCO_2 leads to fall in H_2CO_3 and rise in pH.

Ferric chloride test differentiates metabolic acidosis from diabetic acidosis.

Difficulty in breathing due to ACIDOTIC HYPERPNEA

Hyperventilation (Kussmaul-Kien breathing or air hunger) eliminates carbon dioxide and partially compensates for metabolic acidosis. There is measurable overbreathing at serum bicarbonate of 30 vol.%, and marked overbreathing at 20 vol.%. Pulmonary function is normal.

- Deep, usually rapid breathing.
- Patient may be unaware of overbreathing at first; later becomes dyspneic.
- Dyspnea is persistent, not affected by exertion.
- No paroxysmal nocturnal dyspnea or orthopnea.

ASSOCIATED FINDINGS

- Stupor or coma may be present.
- Lungs normal on auscultation, except for rhonchi in the comatose patient.
- X-ray often normal.

Uremic Acidosis

- Pasty complexion, whitish frost of urea on skin.
- Often nausea and vomiting.
- Sometimes pericardial friction rub.
- Tetany may occur.

Diabetic Acidosis

- Fruity breath odor.
- Soft eyeballs, blurred vision.
- Frequently nausea, vomiting, polyuria and polydipsia.

Other Forms of Acidosis

- Hyperpnea can result from acidosis of severe diarrhea; occurs in adults only with diarrhea of cholera.
- Initial hyperventilation in salicylate poisoning is neurogenic, often absent in children; later metabolic acidosis causes secondary hyperpnea.
- Salicylate poisoning may simulate diabetic acidosis, clinically; differential test by addition of ferric chloride to urine — purple ring appears which persists after boiling.

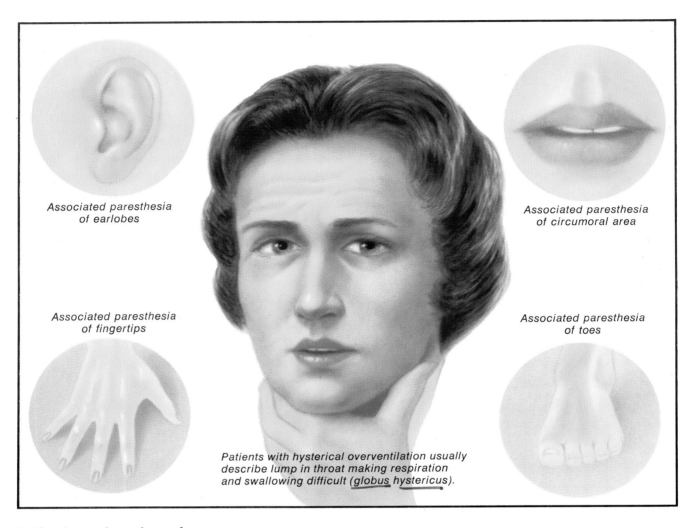

Associated paresthesia of earlobes

Associated paresthesia of circumoral area

Associated paresthesia of fingertips

Associated paresthesia of toes

Patients with hysterical overventilation usually describe lump in throat making respiration and swallowing difficult (globus hystericus).

Difficulty in breathing due to
HYPERVENTILATION SYNDROME

Anxiety is a very common cause of acute and chronic hyperventilation, especially in young women. The condition is often referred to as hysterical overventilation. Neurogenic hyperventilation can also result from organic central nervous system conditions, such as strokes, meningitis, encephalitis and early salicylate poisoning.

- Patient describes chest tightness or inability to take a deep enough breath.
- Dyspnea noted when inactive; may be overlooked during exertion or when patient is diverted.
- Inspiration is deep but not labored.
- Breathing is irregular; frequently interrupted by yawning and sighing.
- No orthopnea.
- Momentary gasping, but no true paroxysmal nocturnal dyspnea.

ASSOCIATED FINDINGS

- Examination of lungs normal.
- Patient may describe lump in throat which seems to make respiration and swallowing difficult — *globus hystericus*.
- Hyperventilation may result in low PCO_2 and alkalosis.
- Sometimes tingling or numbness of fingers, toes, earlobes and circumoral region.
- Palpitation, dizziness, faintness and sharp chest pain may occur.
- Carpopedal spasm and even opisthotonus in some.
- Many symptoms can be duplicated by voluntary overbreathing.
- Can be relieved by breathholding or rebreathing into a paper bag.
- X-ray normal.
- T and, at times, ST changes may be seen on electrocardiogram.

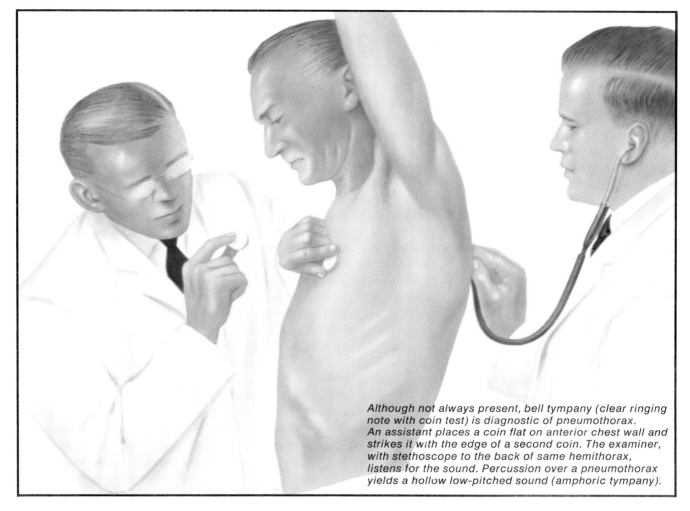

Although not always present, bell tympany (clear ringing note with coin test) is diagnostic of pneumothorax. An assistant places a coin flat on anterior chest wall and strikes it with the edge of a second coin. The examiner, with stethoscope to the back of same hemithorax, listens for the sound. Percussion over a pneumothorax yields a hollow low-pitched sound (amphoric tympany).

Difficulty in breathing due to
PNEUMOTHORAX

A pneumothorax is an accumulation of air or gas in the pleural cavity due to a tear in the pleura, usually over an alveolar cyst but occasionally over a tuberculous cavity. Pneumothorax may also occur with penetrating or nonpenetrating trauma. Signs are detectable when one-third or more of the lung collapses. It occurs most often in men between the ages of 20 and 40.

- Persistent dyspnea.
- Dyspnea not affected by position.
- No paroxysmal nocturnal dyspnea.
- Shallow, rapid breathing; grunting may occur.

ASSOCIATED FINDINGS

- First symptom is pain — sharp and pleuritic; later, persistent and dull.
- Ordinarily no cough.
- Cyanosis in marked cases.
- Large pneumothorax can shift mediastinal contents and cause trachea to deviate to opposite side.
- When extreme (tension pneumothorax with positive pressure in pleural space), hemithorax is large, interspaces wide and percussion note flat.
- Chest motion lags on affected side.
- Lower border of percussion note does not vary with respiration; breath sounds diminished and amphoric or absent; tactile fremitus decreased.
- If stethoscope is placed over posterior chest and coins are tapped over anterior chest, ringing noise is heard.
- Uncommonly, left-sided pneumothorax has clicking sound with each heart beat, audible at a distance.
- X-ray shows peripheral lucency of airspace.
- Collapse varies with respiration; small pneumothorax will be seen only in expiratory x-ray.
- X-ray appearance of underlying lung more likely normal than emphysematous.
- X-ray similar to large bulla; but coin sign is negative with bulla and acute symptoms are not seen.

DIFFERENTIAL DIAGNOSIS OF DIFFICULTY IN BREATHING

	Dyspnea	Orthopnea	Paroxysmal Nocturnal Dyspnea	Character of Breathing
LARYNGEAL OBSTRUCTION	Persistent; usually only on inspiration	None	None	Prolonged, noisy inspiration
PULMONARY EMPHYSEMA	At first exertional; persistent in advanced cases	Most comfortable leaning forward but can lie flat	Usually none	Prolonged expiration; audible wheezing, rhonchi
PNEUMONIA	Persistent	None	None	Rapid and shallow; jerky expiratory grunting in lobar form
PULMONARY FIBROSIS	At first exertional, then persistent	None	None	Rapid, shallow, labored
PULMONARY INFILTRATIVE LESIONS	Usually persistent	None	None	Rapid, shallow, labored
BRONCHIAL ASTHMA	Paroxysmal with acute attacks	Most comfortable leaning forward but can lie flat	Paroxysms can occur when asleep	Prolonged expiration; audible wheezing

Associated Signs and Symptoms	Key Laboratory Data
Usually barking cough; cyanosis may be marked; loud inspiratory stridor, no rales or dullness; inspiratory indrawing of supraclavicular and intercostal spaces; hoarseness	X-ray normal
Usually nonproductive cough, may be productive on awakening; cyanosis may be marked; loud P_2, xiphoid presystolic gallop; fixed chest	X-ray shows hyperlucent lung fields, depressed diaphragm
Severe cough, mucopurulent sputum; cyanosis may be marked; rash in varicella pneumonia; signs of consolidation, marked pleuritic pain and prostration in bacterial lobar pneumonia	X-ray shows diffuse bilateral density in diffuse viral pneumonias; opacification of consolidated lobe in bacterial lobar pneumonia
Usually nonproductive cough; cyanosis usually only on exertion; loud breath sounds, scattered fine rales; poor chest expansion; inspiratory rib space retraction	X-ray shows diffuse linear densities; vital capacity markedly reduced; asbestos bodies in sputum with asbestosis
Usually nonproductive cough; cyanosis may be marked; scattered fine rales; fever common	X-ray shows diffuse patchy densities
Cough — at first nonproductive, then sticky mucoid sputum; cyanosis may be marked; prolonged high-pitched expiratory sounds and wheezes; often sneezing	X-ray shows normal or hyperinflated chest; eosinophils in sputum and blood

(Continued on next page.)

DIFFERENTIAL DIAGNOSIS OF DIFFICULTY IN BREATHING (continued)

	Dyspnea	Orthopnea	Paroxysmal Nocturnal Dyspnea	Character of Breathing
PULMONARY EDEMA OF LEFT VENTRICULAR FAILURE	At first exertional, then persistent	Prominent	Prominent	Rapid, shallow, labored
PULMONARY EDEMA OF MITRAL STENOSIS	Exertional; sometimes persistent	Prominent	Uncommon	Rapid, shallow, labored
OBESITY	Exertional	Present when abdomen very large	None	Rapid, shallow, labored
ACIDOTIC HYPERPNEA	Persistent, not affected by exertion	None	None	Deep, usually rapid
HYPERVENTILATION SYNDROME	Resting; often relieved by exertion	None	Momentary gasping, not true PND	Deep, irregular; with yawning and sighing
PNEUMOTHORAX	Persistent	None	None	Rapid, shallow, sometimes grunting

Associated Signs and Symptoms	Key Laboratory Data
Cough usually nonproductive, may have frothy pink sputum; peripheral cyanosis, no true unsaturation; bilateral inspiratory crepitant rales; apical gallop rhythm	X-ray shows pulmonary vascular congestion, cardiomegaly
Cough usually nonproductive, may have frothy pink sputum; peripheral cyanosis, little true cyanosis; bilateral inspiratory crepitant rales; apical diastolic rumble	X-ray shows pulmonary vascular congestion, left atrial enlargement, enlarged pulmonary arteries
Cough only with associated bronchitis; cyanosis may be marked; often xiphoid presystolic gallop and loud P_2	X-ray usually normal
Often stupor or coma	X-ray often normal; ferric chloride test to differentiate salicylate and diabetic acidosis
Sometimes tingling of toes, fingers, circumoral area and earlobes; often *globus hystericus*	X-ray normal; sometimes T and ST changes on ECG
Cyanosis may be marked; decreased breath sounds, dullness and hyperresonance on affected side; positive coin sign; tracheal deviation	X-ray shows lucent, avascular area in peripheral lung

MIDLINE CHEST PAIN

Painful stimuli from organs within the chest (heart, lungs, pericardium, esophagus) localize to the middle of the anterior chest wall, since impulses from the autonomic fibers supplying these organs are all referred to this somatic site.

Referred pain, as a rule, is imprecise in location and often difficult to describe; therefore, there is a considerable overlap in midline chest pain caused by various disease states. Local conditions of the chest wall can also produce midline chest pain and further complicate diagnosis.

Nonetheless, careful analysis of the pain as to character, duration, intensity, location, radiation, precipitating factors, methods of relief, as well as associated findings, can facilitate correct diagnosis.

Thirteen major diagnostically important causes of midline chest pain are described individually and then compared in tabular form.

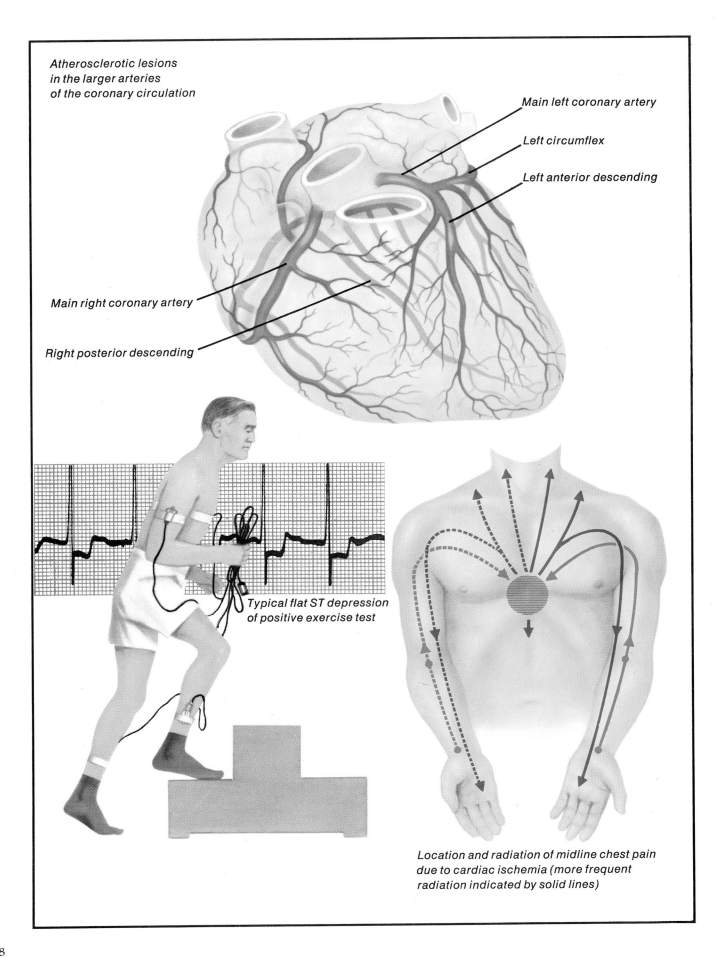

Midline chest pain due to
ANGINA PECTORIS

Literally a strangling of the chest, the term angina pectoris has come to mean the brief intermittent pain of ischemic heart disease. Similar chest pain of aortic valve disease is also referred to as angina pectoris.

- Pain is pressing, squeezing, strangling, crushing; or aching with tightness and heaviness; occasionally burning; almost never sticking or jabbing.
- Most often substernal; rarely precordial.
- May radiate to either shoulder or to ulnar surface of either arm; to neck, jaw, teeth or epigastrium; occasionally to interscapular region of back; can begin in wrist or elbow and radiate to chest; can localize in site of other pain — in jaw with trigeminal neuralgia, in shoulder with bursitis.
- Lasts 1 to 3 minutes, sometimes up to 15 minutes.
- Can be precipitated by emotional stress; most often occurs during exertion, especially in morning; more likely with unaccustomed activity.
- Occurs with less exertion after meals; may occur only after meals in the sedentary.
- More likely in cold weather, inhaling cold air.
- Pain *after* exertion or pain reproduced by hyperventilation, pressure on chest or movement of shoulder is usually not angina pectoris.
- May be partially relieved by belching — simulating G.I. distress; promptly relieved by nitroglycerin.

ASSOCIATED FINDINGS

- Fear and anxiety; pale complexion; sometimes sweating.
- No fall in blood pressure; no pulmonary edema.
- Apical presystolic gallop and palpable atrial kick may be present during pain; ECG often demonstrates ST shifts during pain, normal between attacks in more than half.

Midline chest pain due to
CORONARY INSUFFICIENCY

The term coronary insufficiency, at times used loosely, is being replaced by the term "preinfarction angina," properly referring to the intermediate syndrome of myocardial ischemia caused by coronary atherosclerotic disease. The pain of coronary insufficiency is more severe than that of angina pectoris. The cardiac muscle necrosis characterizing myocardial infarction is absent in coronary insufficiency. The condition may be prodromal to myocardial infarction.

- Character and distribution of pain similar to that of angina pectoris; patient may describe pain as more severe than previous angina.
- Duration from a few minutes to several hours; may recur for days, occasionally weeks.
- Frequently occurs at rest without apparent trigger.
- May be accompanied by marked increase in frequency of typical anginal attacks.
- Relieved most often by narcotic analgesics.

ASSOCIATED FINDINGS

- Usually fear and anxiety; ashen complexion; often profuse sweating.
- Unlike myocardial infarction, fever and SGOT elevation not usual; SGOT elevation, however, may be detected within first 7 to 14 days, if infarct develops.
- Dyspnea, labored rapid breathing, apical gallop rhythm and bilateral basal crepitant rales of left ventricular failure may be present.
- QRS changes absent; ST and T abnormalities do not persist more than 48 hours after pain — unlike myocardial infarction.

Midline chest pain due to
MYOCARDIAL INFARCTION

Myocardial infarction, a clinical syndrome produced by acute ischemic necrosis of the myocardium, is ordinarily due to coronary atherosclerosis and often thrombosis. It is especially prevalent in persons with diabetes, hypertension, hypercholesterolemia and hyperuricemia, who are prone to coronary disease. Coronary insufficiency or a sudden increase in the frequency of angina is prodromal in at least one-third.

- Character and distribution of pain similar to angina pectoris, but more severe.
- Patient may be unable to describe pain; instead may move palm over sternum or slowly clench fist.
- Persistent; unaffected by position.
- Occurs mostly at rest; occasionally on exertion.
- Duration of less than an hour to several days.
- Relieved only by narcotic analgesics.

ASSOCIATED FINDINGS

- Fear and anxiety.
- Often ashen complexion; profuse sweating.
- Nausea, vomiting and belching in many patients; hiccoughs frequent, may be very distressing.
- Fever may appear 4 to 8 hours after onset of pain.
- Dyspnea, labored rapid breathing, apical gallop rhythm and bilateral basal crepitant rales of left ventricular failure may be present.
- Thready pulse, low or absent blood pressure, falling urine output of shock syndrome may be present.
- Irregular pulse of arrhythmia in most, but not always on initial examination.
- Soft first heart sound and presystolic gallop even without gross heart failure; pericardial friction rub in some, may not occur for a day or two after infarction.
- ECG changes and elevation of SGOT will establish diagnosis; may not be evident for several days.

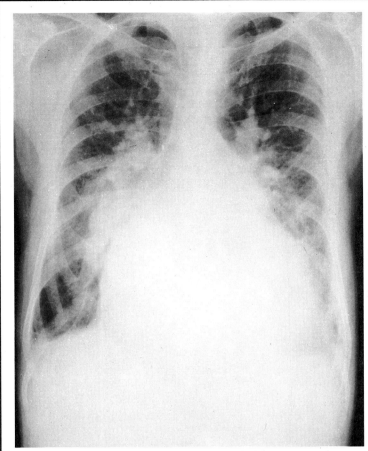

Cardiomegaly and large pulmonary arteries typical of pulmonary hypertension, in this case secondary to atrial septal defect

Location and radiation of midline chest pain due to pulmonary hypertension

Midline chest pain due to
PULMONARY HYPERTENSION

Severe and chronic pulmonary hypertension, caused by rheumatic mitral stenosis, congenital heart disease or primary pulmonary hypertension, can produce chest pain similar to that of cardiac ischemia. Whether it actually stems from the heart or from the pulmonary vascular bed is uncertain. Chest pain can also occur in pulmonic stenosis with no pulmonary hypertension.

- Character and distribution of pain similar to that of angina pectoris — pressing, crushing, bursting.
- May be brief as in angina pectoris, or prolonged as in myocardial infarction.
- At times worsened by breathing, unlike angina pectoris.
- Often occurs upon exertion.
- Will occur at times with anoxia.
- Sometimes relieved by oxygen.

ASSOCIATED FINDINGS

- Fear and anxiety.
- Syncope may occur, especially in primary pulmonary hypertension.
- Patient often dyspneic from underlying pulmonary and/or heart disease.
- Cyanosis common with interventricular or interatrial septal defect, or with patent ductus arteriosus when shunt is from right to left.
- Loud P_2, except in pulmonic stenosis.
- Right ventricular heave and/or presystolic gallop.
- Systolic murmur of tricuspid insufficiency or diastolic murmur of pulmonic insufficiency may be present.
- Chest x-ray will always show large main pulmonary arteries and obliterated distal pulmonary vasculature.
- ECG shows right ventricular hypertrophy and P pulmonale.

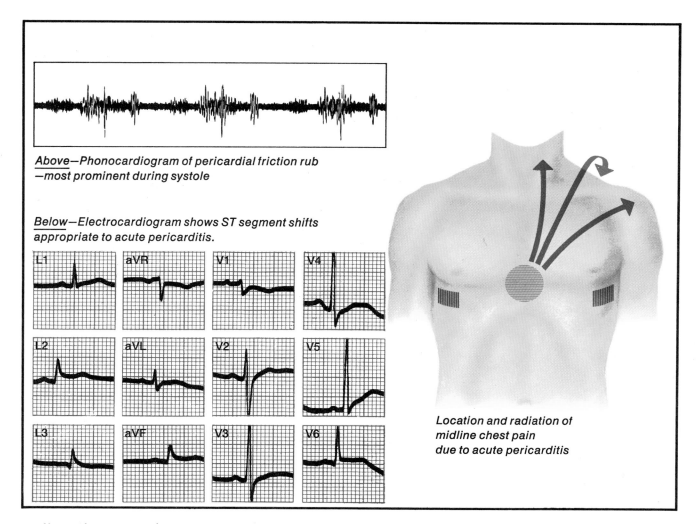

Above—Phonocardiogram of pericardial friction rub —most prominent during systole

Below—Electrocardiogram shows ST segment shifts appropriate to acute pericarditis.

Location and radiation of midline chest pain due to acute pericarditis

Midline chest pain due to
ACUTE PERICARDITIS

In the past, tuberculosis and rheumatic fever were frequent causes of acute pericarditis. Today, the cause is usually unknown, probably most often a viral infection. Acute pericarditis is frequently preceded by an upper respiratory infection, with pain as the first symptom. Chest pain is usually not a feature of chronic pericarditis or pericarditis of metastatic carcinoma. The main differential is with myocardial infarction.

- Sharp, as pleural pain; can be constricting, pressing and aching, as in myocardial infarction.
- Usually lower substernal, but can be in either right or left side of anterior chest.
- May radiate to left shoulder, neck or back; rarely to jaw or below elbow as in myocardial infarction.
- Onset at rest; continuous or intermittent; lasting for days or weeks — longer than in myocardial infarction.
- Accentuated by breathing in most, by lying prone or by swallowing in some.
- Relieved only by narcotic analgesics.

ASSOCIATED FINDINGS

- Some associated pleuritis in nearly every case; nonproductive cough may be present.
- Pericardial friction rub heard during systole and diastole (only in systole in one-third).
- Unlike myocardial infarction, rub begins with pain and often lasts as long as pain.
- Fever always present; lasts about a week in idiopathic, longer in tuberculous and rheumatic.
- Soft heart tones and paradoxical pulse (fall in systolic pressure of more than 10 mm Hg with inspiration) when pericardial effusion is present.
- Low arterial blood pressure and distended neck veins, caused by large and rapid accumulation of pericardial fluid (cardiac tamponade).
- ECG may show ST elevation followed by T inversion in all standard bipolar limb and precordial leads.
- Regular alternation in amplitude of QRS (electrical alternans) frequently present.

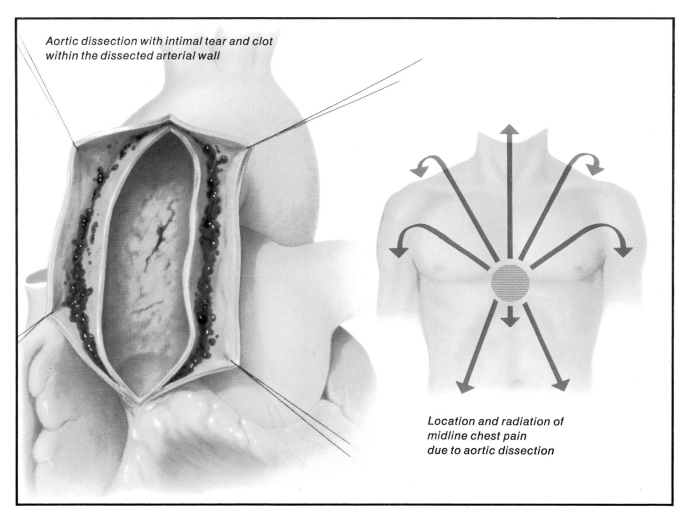

Aortic dissection with intimal tear and clot within the dissected arterial wall

Location and radiation of midline chest pain due to aortic dissection

Midline chest pain due to AORTIC DISSECTION

Aortic dissection may occur as a complication of hypertension in the older population. It is sometimes seen in the young as a result of isolated cystic medial necrosis, as a part of Marfan's syndrome or as a complication of coarctation of the aorta. Leakage from an aortic dissection may lead to bloody pleural effusion. Pain is the initial and most prominent symptom.

- Pain is severe; tearing or bursting.

- Substernal; radiates to back; may spread to head, extremities or abdomen, depending on pattern of dissection.

- Occurs at rest, sometimes on exertion; persistent, unaffected by position; relieved only by narcotic analgesics.

ASSOCIATED FINDINGS

- Dyspnea frequent, breathing not labored.

- Ashen complexion; profuse sweating.

- Blood pressure often surprisingly high; fever develops after a few days.

- Confusion and disorientation with carotid dissection.

- Paresthesias, hemiplegia, aphasia and blindness with extension to vessels of head.

- Gross hematuria if renal arteries are involved.

- Cold, clammy, mottled leg and absent pulse with extension to femoral artery.

- Marked differences in arterial pulses; bisferiens carotid pulse; sudden appearance of palpable aortic pulsation in sternoclavicular region.

- Cardiomegaly from pre-existing hypertension.

- Appearance of aortic diastolic murmur indicates distention of aortic ring.

- Marked leukocytosis common (20,000-30,000).

- ECG normal or shows minor ST and T changes or left ventricular hypertrophy due to hypertension.

- Chest x-rays show widened aorta and may reveal diagnostic double border.

- Appearance of pericardial friction rub of early hemopericardium may simulate pericarditis.

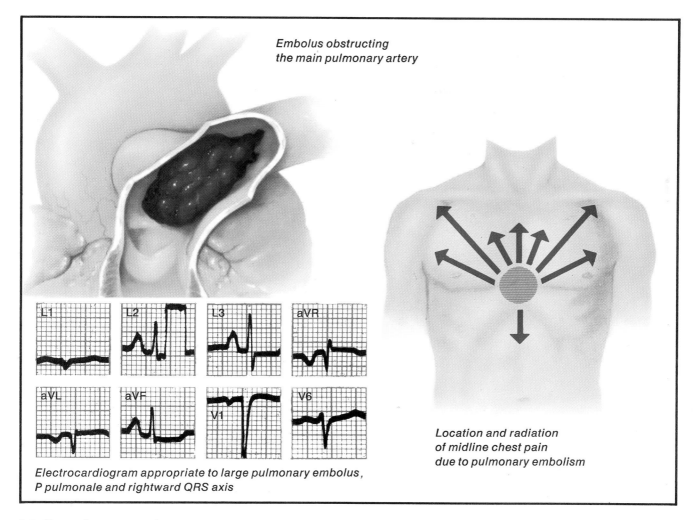

Embolus obstructing the main pulmonary artery

Electrocardiogram appropriate to large pulmonary embolus, P pulmonale and rightward QRS axis

Location and radiation of midline chest pain due to pulmonary embolism

Midline chest pain due to
PULMONARY EMBOLUS

Embolization of a thrombus to the lung from the right atrium, pelvis or lower extremities can occur in patients immobilized or at bed rest. Pulmonary embolus can also occur in congestive heart failure, venous insufficiency, polycythemia and sickle cell disease. Emboli to the main pulmonary arteries usually cause midline chest pain; smaller emboli do not. A single massive embolus may cause sudden death.

- Pain oppressive; sudden in onset; substernal.
- May radiate elsewhere in chest, shoulders and epigastrium; rarely to arms or neck.
- Onset at rest; lasts for minutes to hours.
- Relieved only by narcotic analgesics.

ASSOCIATED FINDINGS

- Dyspnea often marked.
- Breathing rapid and shallow; normal breath sounds.
- Ashen complexion; profuse sweating.
- Cyanosis in severe cases.
- Fever appears in a few hours.
- Transient elevations in leukocyte count and sedimentation rate may occur.
- Jaundice can develop later.
- Blood pressure often falls to shock levels.
- Pulse rapid; often irregular due to atrial fibrillation.
- Loud P_2 and xiphoid presystolic gallop from pulmonary hypertension.
- Distended neck veins and hepatomegaly (acute right ventricular failure) often follow.
- Hemoptysis, pleuritic pain and pleural friction rub (signs of pulmonary infarct) develop later.
- ECG often shows P pulmonale, shift in QRS axis to right and T inversion in leads 2, 3, aVF, V1 and V2.
- Right bundle branch block may occur transiently.

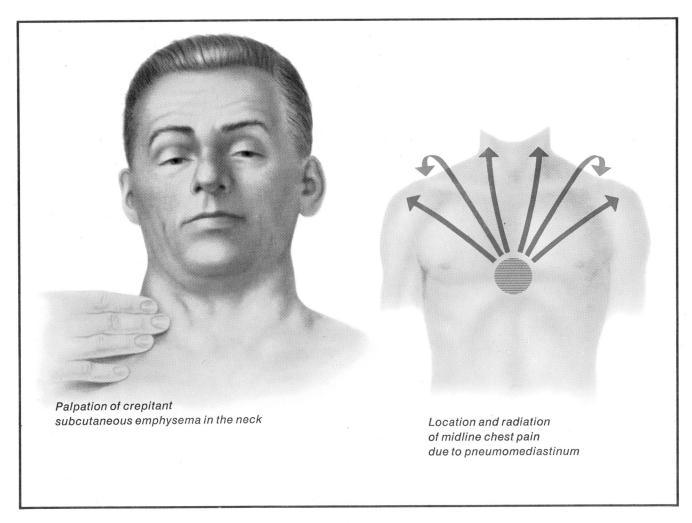

Palpation of crepitant subcutaneous emphysema in the neck

Location and radiation of midline chest pain due to pneumomediastinum

Midline chest pain due to
PNEUMOMEDIASTINUM

Pneumomediastinum usually results from leakage of air from interstitial spaces of the lung to the mediastinum because of ruptured alveoli. It can be caused by heavy lifting, Valsalva's maneuver, coughing, asthma or excessive positive pressure during anesthesia. Pneumomediastinum may also follow perforation of the trachea, bronchi or esophagus, or even rupture of a hollow viscus in the abdomen.

- Pain severe and sudden, usually described as pressing.
- Pain may be only finding in small pneumomediastinum.
- Predominantly substernal.
- May radiate to back, neck and shoulders; rarely to arms.
- Persistent; lasts hours to days.
- Relieved only by narcotic analgesics.

ASSOCIATED FINDINGS

- No dyspnea except in severe cases with positive pressure in mediastinum.
- Fever not present except from underlying cause, *e.g.*, ruptured viscus.
- Typical crunching sound heard over heart with each beat; auscultation of heart and lungs otherwise normal.
- When air spreads from mediastinum to neck and body, skin becomes swollen and crepitant to palpation (subcutaneous emphysema).
- Air usually leaks from mediastinum to left pleural space, causing small left pneumothorax.
- Air occasionally leaks into retroperitoneal space, causing abdominal swelling.
- Rarely, neck veins become distended, with associated dyspnea, cyanosis and shock, when positive pressure in mediastinum blocks venous return; surgical intervention may be necessary.
- Air in mediastinum and small left pneumothorax best seen along left heart border in expiratory chest x-ray.

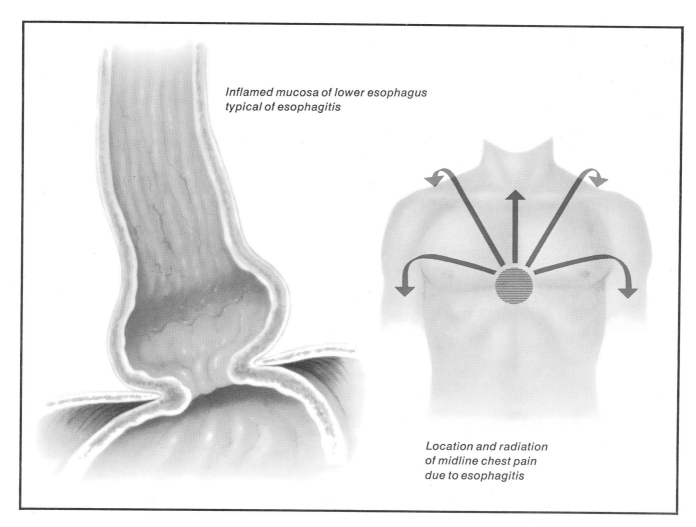

Inflamed mucosa of lower esophagus typical of esophagitis

Location and radiation of midline chest pain due to esophagitis

Midline chest pain due to ESOPHAGITIS

Esophagitis, a painful inflammation of the lower third of the esophagus, is often associated with peptic disease or with hiatus hernia. It can also result from prolonged intubation following surgery or from ingestion of such irritants as lye.

- Pain is burning, gripping, squeezing; may be as severe as in cardiac ischemia.
- Primarily substernal.
- May radiate to neck, back, shoulders and arms.
- May last few minutes or hours.
- May be triggered by swallowing or by lying recumbent.
- May appear spontaneously.
- Often follows heavy meals, ingestion of spices or alcohol.
- Usually relieved by antacids, at times by nitroglycerin.
- Can often be duplicated by perfusion of acid into lower esophagus — not reliable as diagnostic test.

ASSOCIATED FINDINGS

- Heartburn.
- Nausea, vomiting.
- Excessive salivation.
- Transitory difficulty in swallowing (reflex esophagospasm).
- Persistent dysphagia can be caused by scarring of lower esophagus in chronic esophagitis.
- Sometimes hematemesis and melena from mucosal erosion and bleeding.
- Anemia from chronic blood loss.
- Heart and lungs normal on examination.
- Pain may trigger heart block, extrasystoles, angina, asthma or laryngospasm in susceptible individuals.
- Definitive diagnosis by esophagoscopy.

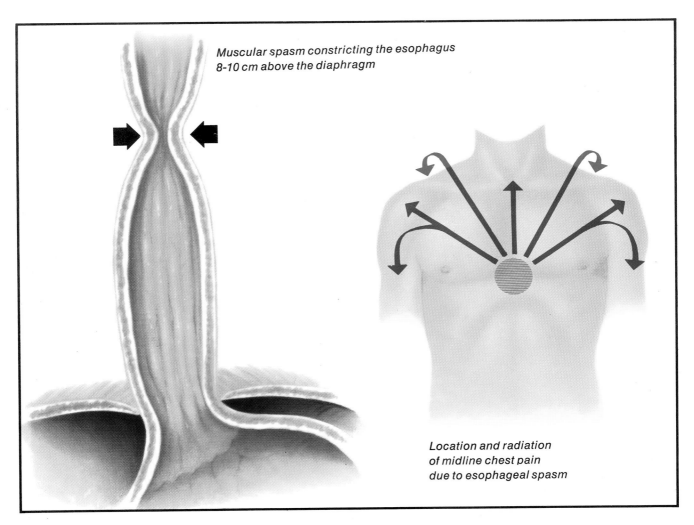

Midline chest pain due to
ESOPHAGEAL SPASM

Esophageal spasm is a transitory painful contraction, often ring-like, of a portion of the lower esophagus. It is frequently confused with cardiospasm (achalasia), in which the lower end of the esophagus fails to relax while the rest of the esophagus is markedly dilated. Chest pain, however, is seldom present in achalasia.

- Pain may be pressing or squeezing, as in cardiac pain.
- Usually substernal.
- May radiate to neck, back, shoulders and arms.
- Lasts for moments or hours.
- Usually triggered by swallowing (especially rough food).
- Pain persists.
- Pain may follow emotional distress (like angina); never provoked by exertion (unlike angina).
- Sometimes relieved by nitroglycerin.
- Usually not relieved by antacids.

ASSOCIATED FINDINGS

- Transitory difficulty in swallowing.
- Bleeding and blood-loss anemia do not occur as in esophagitis.
- Heart normal on examination.
- Lungs normal on examination.
- X-ray of esophagus may be normal or show ring-like constriction with dilatation.
- Spasm best demonstrated by fluoroscopy after patient swallows irritant, such as hard bread.
- On fluoroscopy, constriction seen 8 to 10 cm above diaphragm.

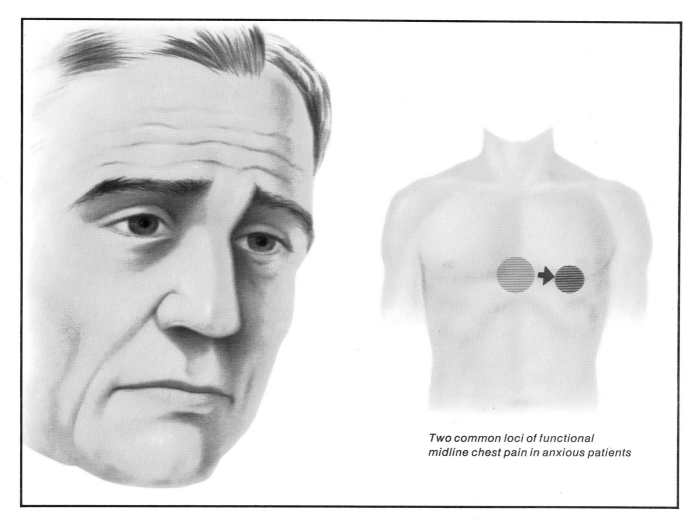

Two common loci of functional midline chest pain in anxious patients

Midline chest pain due to
ANXIETY

One of the most important differentials in the diagnosis of cardiac chest pain is anxiety-induced pain. Pain resulting from other causes can be aggravated or accentuated by anxiety; and anxiety may often complicate the interpretation of pain. Descriptions of pain by anxious patients are as variable as their imaginations; however, certain patterns are common.

- Pain may be described as repeated, sharp and jabbing, lasting only seconds, or as dull, aching and almost continuous.
- Anxious patients in marked distress may describe pain as very severe.
- Usually localized to precordial area.
- Rather than pain, some patients may complain of intermittent substernal tightness with typical symptoms of hyperventilation.
- Frequently precipitated by emotional stress or phobia of overcrowding.
- Almost never precipitated by exertion.
- Radiation variable.
- Usually relieved by physician reassurance.

ASSOCIATED FINDINGS

- Typical anxious appearance.
- Patient often convinced of heart disease before seeing physician.
- Occasional extrasystoles and variable chest wall tenderness only abnormal physical findings.
- ST and T abnormalities may be present because of associated hyperventilation.
- Frequently breathlessness and sighing.
- Palpitation and dizziness of hyperventilation often present.
- Coronary arteriography may be necessary to clarify diagnosis in some patients.

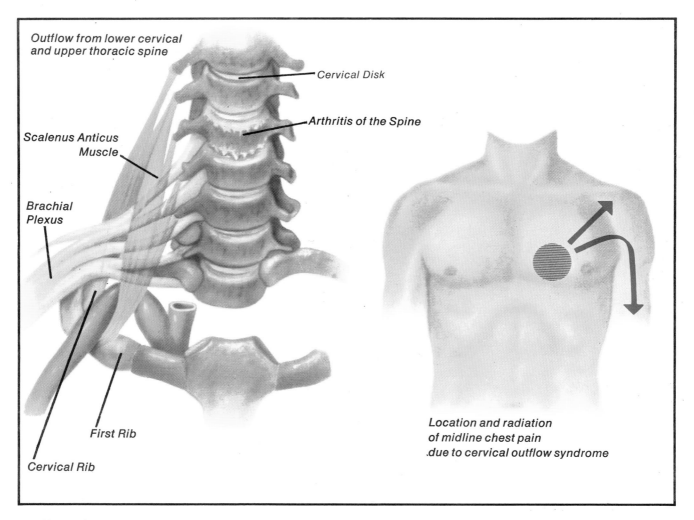

Location and radiation of midline chest pain due to cervical outflow syndrome

Midline chest pain due to
CERVICAL OUTFLOW SYNDROME

A cervical outflow syndrome may cause midline chest pain by nerve irritation or compression at the outflow from the lower cervical and upper thoracic spine, as in osteoarthritis, kyphosis or ruptured intervertebral disk. The condition may also arise from irritation or compression at the brachial plexus by the cervical rib or the scalenus anticus muscle, or by the first rib and clavicle.

- Pain usually sharp and lancinating; may be dull and aching.
- Anterior chest, left shoulder and arm.
- In early stages, localized by dermatome distribution of nerves; later becomes more diffuse.
- Prolonged or intermittent.
- Onset follows movement of neck with compression at spine, or movement of arm with compression at brachial plexus.

ASSOCIATED FINDINGS

- Sensory changes in index and middle fingers with unilateral rupture of 6th intervertebral disk; in fourth or fifth finger with compression at brachial plexus.
- Involved area cold and clammy.
- Affected muscles or spine usually tender to touch.
- When compression is at spine, anvil test may be positive — blow on top of head compresses nerves and reproduces pain.
- Pain from compression at spine relieved by cervical traction.
- When compression is at brachial plexus, Adson test often positive — radial pulse altered or obliterated when patient sits with hands on knees, takes a deep breath, elevates and turns chin to affected side.
- With brachial plexus compression, movement of arm through full range will reproduce pain.
- Scalenus anticus syndrome confirmed by relief of pain with procaine injection into muscle.
- In costoclavicular syndrome, radial pulse disappears and pain is reproduced by pushing shoulder down and back, thereby compressing brachial plexus between first rib and clavicle.

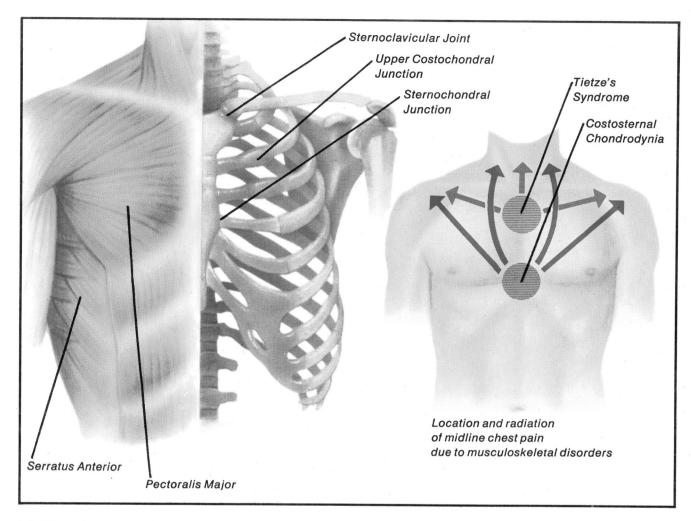

Location and radiation of midline chest pain due to musculoskeletal disorders

Midline chest pain due to
MUSCULOSKELETAL DISORDERS

Midline chest pain may be produced by several types of musculoskeletal disorders. For example, myositis of various chest muscles, Tietze's syndrome or costosternal chondrodynia may be implicated in this type of pain. Tietze's syndrome usually is seen in patients under 40 years of age, whereas costosternal chondrodynia is more often seen in patients over 40.

- Pain usually dull and pressing.
- Often ill-defined with myositis.
- Generally not severe with Tietze's syndrome or costosternal chondrodynia.
- Upper substernal in Tietze's syndrome.
- Lower substernal in costosternal chondrodynia.
- In anterior chest in myositis, not precisely localized to involved muscle.
- May radiate to shoulder and neck in Tietze's syndrome and costosternal chondrodynia.
- May radiate down to ulnar surface of arm when pectoralis major is involved.
- Pain usually prolonged, hours to days; sometimes recurrent.
- With involvement of pectoralis minor, pain can be reproduced by pushing shoulder down and forward against resistance.
- Pain usually aggravated by movement of chest wall, often worsened by lying down or by damp weather — in Tietze's syndrome and costosternal chondrodynia.
- In myositis, trigger point often found in muscle; pressure will reproduce pain.

ASSOCIATED FINDINGS

- In Tietze's syndrome, tenderness and distinct swelling of sternoclavicular joint or one of upper costosternal joints.
- In costosternal chondrodynia, tenderness without swelling of several costosternal junctions from third to fifth rib.
- In myositis, increased sweating over involved muscle; tenderness on palpation of involved muscle.

DIFFERENTIAL DIAGNOSIS OF MIDLINE CHEST PAIN

	Character	Location	Radiation	Duration
ANGINA PECTORIS	Pressing, crushing, bursting	Substernal	Shoulders, arms, neck, jaw	Usually 1 to 3 minutes
CORONARY INSUFFICIENCY	Pressing, crushing, bursting	Substernal	Shoulders, arms, neck, jaw	Few minutes to several hours
MYOCARDIAL INFARCTION	Pressing, crushing, bursting	Substernal	Shoulders, arms, neck, jaw	Less than an hour to several days
PULMONARY HYPERTENSION	Pressing, crushing, bursting	Substernal	Shoulders, arms, neck, jaw	Brief, like angina; or long, like infarct
ACUTE PERICARDITIS	Crushing or sharp	Substernal	Shoulders, neck, or back (not arms or jaws)	Days to weeks
AORTIC DISSECTION	Tearing or bursting	Substernal	Back, head, extremities, abdomen	Minutes to hours, may recur
PULMONARY EMBOLUS	Sudden and oppressive	Substernal	Chest, shoulders, epigastrium	Minutes to hours, may recur
PNEUMOMEDIASTINUM	Sudden and severe	Substernal	Back, neck, shoulders	Hours to days

Onset	Relieved by	Associated Signs and Symptoms	Key Laboratory Data
Exertion; emotional stress	Nitroglycerin	Fear and anxiety; patient pale, sometimes sweating; may have apical presystolic gallop and atrial kick	ECG may show ST depression during exercise
Usually at rest	Most often by narcotic analgesics	Fear and anxiety; dyspnea; patient pale and often sweating; may have gallop and rales	ECG may show ST and T changes; temperature normal; SGOT elevation may be detected within first 7-14 days if infarct develops
Usually at rest	Narcotic analgesics only	Fear and anxiety; dyspnea; weakness; shock; arrhythmia; acute left ventricular failure	Q waves and other acute changes on ECG; SGOT and temperature elevated
Exertion; increasing anoxia	Sometimes by oxygen	Fear and anxiety; dyspnea; sometimes syncope; cyanosis; loud P_2; right ventricular heave and gallop	Chest x-ray shows large main pulmonary arteries; ECG — right ventricular hypertrophy
Usually at rest	Narcotic analgesics only	Fever; nonproductive cough; pericardial friction rub; paradoxical pulse in more severe	ST elevation and T inversion in bipolar limb and precordial leads, electrical alternans
Usually at rest, sometimes on exertion	Narcotic analgesics only	Dyspnea; neurological signs; hematuria; ashen, sweating; differences in pulses; blood pressure often high; bisferiens carotid pulse	ECG normal, or minor changes of left ventricular hypertrophy; x-ray — widening of aorta
Usually at rest	Narcotic analgesics only	Marked dyspnea, rapid shallow breathing; often cyanotic; low blood pressure; loud P_2; hemoptysis and pleural pain later	Shift of QRS axis to right; transitory right bundle branch block
Coughing; sneezing; straining at stool	Narcotic analgesics only	Dyspnea in severe cases; crunching sound over heart with each beat; subcutaneous emphysema	Air seen along left heart border in expiratory chest x-ray

(Continued on next page.)

DIFFERENTIAL DIAGNOSIS OF MIDLINE CHEST PAIN (continued)

	Character	Location	Radiation	Duration
ESOPHAGITIS	Burning, gripping, squeezing	Substernal	Back, neck, shoulders, arms	Minutes to hours
ESOPHAGEAL SPASM	Pressing, squeezing	Substernal	Back, neck, shoulders, arms	Seconds to hours
ANXIETY	Sharp and jabbing, or dull and aching	Often precordial; substernal tightness	Variable	Seconds to continuous
CERVICAL OUTFLOW SYNDROME				
Spinal	Sharp and lancinating, or dull and aching	Anterior chest	Shoulder, arm, neck	Minutes to nearly continuous
Brachial Plexus	Sharp and lancinating, or dull and aching	Anterior chest	Shoulder, arm, neck	Minutes to nearly continuous
MUSCULOSKELETAL DISORDERS				
Tietze's Syndrome	Dull and pressing	Upper substernal	Shoulder and neck	Days to weeks
Costosternal Chondrodynia	Dull and pressing	Lower substernal	Shoulder and neck	Days to weeks
Myositis	Dull and pressing	Anterior chest	Ulnar surface of arm	Prolonged

Onset	Relieved by	Associated Signs and Symptoms	Key Laboratory Data
Swallowing; lying recumbent; after alcohol or spices	Antacids; sometimes nitroglycerin	Heartburn, nausea, vomiting, dysphagia; hematemesis; extrasystoles or heart block during pain in some; acid perfusion may reproduce pain	Anemia when associated with chronic or massive blood loss
Swallowing (especially rough food); emotional stress	Sometimes nitroglycerin; not usually by antacids	Occasional dysphagia	X-ray may be normal; best demonstrated by fluoroscopy; no anemia
Emotional stress; rarely by exertion	Tranquilizers and reassurance	Breathlessness; palpitation; dizziness; frequent sighing; anxious appearance; muscle tenderness	ST and T changes with hyperventilation
Movement of neck	Cervical traction	Painful area cold and clammy; muscle tenderness and tenderness over spine; anvil test often positive	—
Movement of arm or shoulder	Analgesics; infiltration of scalenus anticus	Painful area cold and clammy; muscle tenderness; Adson test often positive; pain may be reproduced by pushing shoulder down and back	—
Movement of chest wall; damp weather	Analgesics	Tenderness with swelling of usually one sternoclavicular joint or costochondral junction	—
Movement of chest wall; damp weather	Analgesics	Tenderness without swelling of several costosternal junctions	—
Can occur at any time	Analgesics	Increased sweating over affected muscle; muscle tender to palpation, may have trigger point; pain may be reproduced by pushing shoulder down and forward against resistance (pectoralis minor)	—

CHEST PAIN ON BREATHING

Conditions involving the pleurae, diaphragm, intercostal muscles and rib cage usually produce chest pain — either only on breathing or worsened by breathing. This symptom is most valuable in diagnosis as it can be accurately described by the patient, and can be closely related to the anatomic site of the reason for the pain. Rapid shallow breathing is often an associated sign.

Many conditions produce chest pain on breathing. A careful analysis of the location of the pain and associated signs and symptoms is necessary to make a precise diagnosis.

Twelve important causes of chest pain on breathing are described individually and then compared in tabular form.

Diagram of typical sound recording of pleural friction rub showing inspiratory and expiratory components

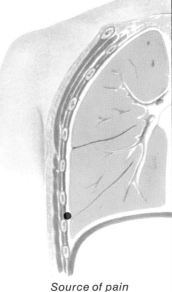

Source of pain on breathing in acute pleurisy

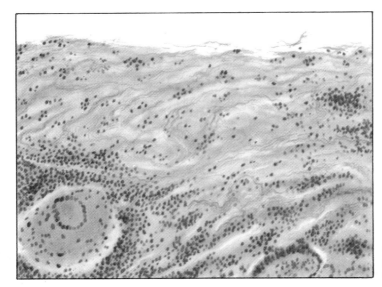

Microscopic appearance of pleural lesion of tuberculosis showing noncaseating granuloma.

Chest pain on breathing due to
ACUTE PLEURISY

Acute pleurisy, an acute inflammation of the pleura without involvement of the underlying lung, can be caused by an inconsequential viral infection. Usually, however, it is a sign of tuberculosis, collagen disease or a neoplasm. It may occur with or without pleural effusion.

- Pain sharp and jabbing only on breathing.
- Markedly intensified by deep breathing, movement of chest, coughing or sneezing.
- Usually localized to lower lateral chest.
- May radiate to back, shoulders and arms; occasionally to abdomen.
- Onset acute.
- Occasionally may precede effusion by weeks.
- Disappears when sufficient effusion accumulates.

ASSOCIATED FINDINGS

- Initial physical findings — palpable friction fremitus; breathy "to and fro" inspiratory and expiratory pleural friction rub.
- Breathing rapid and shallow; proportional to pain.
- Fever nearly always present.
- Nonproductive cough.
- As effusion develops, the involved side reveals flatness to percussion, absent tactile fremitus and obscured breath sounds at lung base.
- Occasionally very loud breath sounds can be produced by fluid compressing lung.
- Weight loss often associated.
- In tuberculous pleurisy, effusion is almost always present; tuberculin test positive, sputum and gastric cultures may be negative.
- In rheumatic fever, rheumatoid arthritis and systemic lupus erythematosus, pleurisy is never an isolated finding; other features significant in diagnosis.
- In subpleural bronchogenic carcinoma, pleurisy usually produces effusion high in LDH.

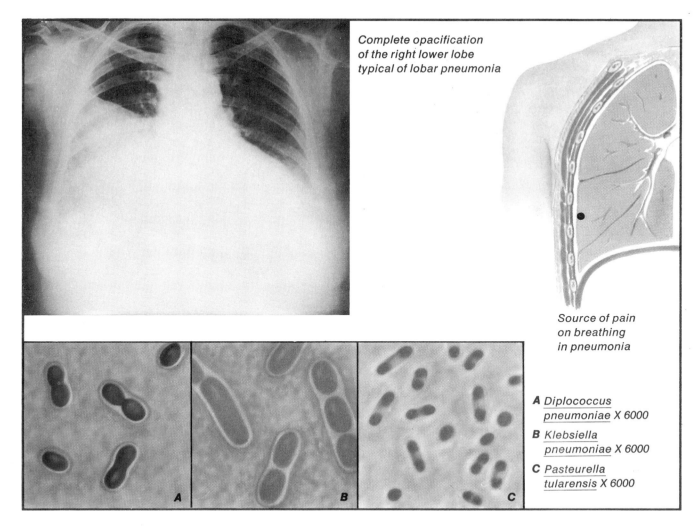

Complete opacification of the right lower lobe typical of lobar pneumonia

Source of pain on breathing in pneumonia

A *Diplococcus pneumoniae* X 6000
B *Klebsiella pneumoniae* X 6000
C *Pasteurella tularensis* X 6000

Chest pain on breathing due to
PNEUMONIA

Pleurisy associated with some types of pneumonia can cause chest pain on breathing. It is prominent in bacterial lobar pneumonia, though not a feature of diffuse viral pneumonia. The pain is more marked with pneumonia of the lower lobes.

- Pain sharp and jabbing only on breathing.
- Markedly intensified by deep breathing, movement of chest, sneezing or coughing.
- Localized over involved lobe; usually lower lateral chest.
- Onset acute in most; preceded by chills in many.
- Persists for several days.
- Lying on involved side reduces pain.

ASSOCIATED FINDINGS

- Severe dyspnea often present, even at rest.
- Severe hacking cough.
- Sputum purulent ("rusty," "red currant jelly" or "sticky").
- Skin warm and flushed.
- Fever and prostration; temperature often rises quickly to 104°F. or 105°F.
- Respiration rapid, jerky and shallow with expiratory grunting.
- Cyanosis variable.
- Signs of consolidation over involved lobe.
- Fine and medium inspiratory rales.
- Rales become coarse and numerous during resolution.
- Pleural friction rub.
- WBC elevated; increase in polymorphonuclear leukocytes.
- X-ray shows opacification of consolidated lobe.
- Tularemia should be suspected when chest pain on breathing, signs of pneumonia and enlarged axillary nodes occur in a hunter, butcher or veterinarian.

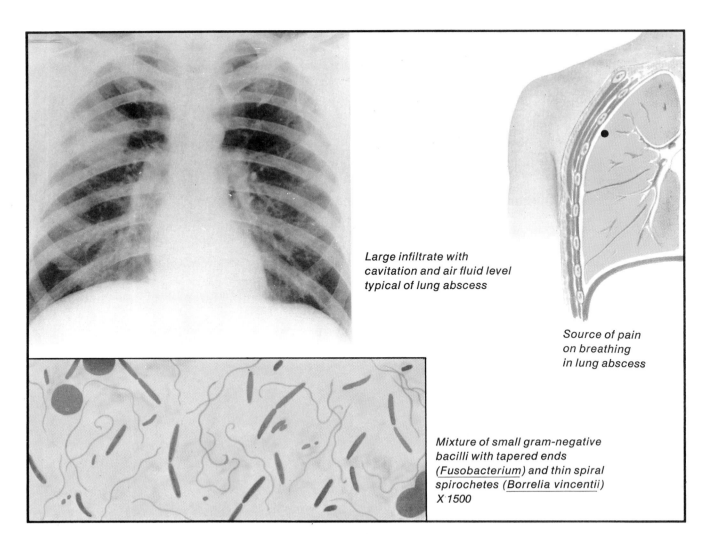

Large infiltrate with cavitation and air fluid level typical of lung abscess

Source of pain on breathing in lung abscess

Mixture of small gram-negative bacilli with tapered ends (Fusobacterium) and thin spiral spirochetes (Borrelia vincentii) X 1500

Chest pain on breathing due to
LUNG ABSCESS

A suppurative focus with necrosis and cavitation, frequently subpleural, can result from septicemia, transdiaphragmatic spread, infection of a hematoma, or penetrating trauma. Most cases are bronchogenic in origin. Obstruction to bronchial drainage is an important cause. Twenty-five percent are caused by bronchostenosis of bronchogenic carcinoma. Lung abscess may also follow aspiration of stomach contents during unconsciousness or aspiration of septic material during tonsillectomy or dental extraction.

- Pain usually sharp and jabbing, at times dull and aching; occurs only on breathing; markedly intensified by deep breathing, movement of chest, sneezing or coughing.
- Localized in chest wall over abscess; no radiation.
- Persists only during acute stage.
- Excruciating with dyspnea when abscess ruptures into pleural space (pyopneumothorax).

ASSOCIATED FINDINGS
- Preceded by nonproductive cough, fever, malaise.
- Cough becomes productive with copious sputum; often fetid and/or bloody with spirochete and fusiform organisms.
- Breathing shallow, proportional to pain.
- Flushed face, fever.
- Dullness to percussion and crepitant inspiratory rales where pain is marked; amphoric breath sounds, as heard in a chronic tuberculous cavity, not heard unless bronchial communication has been established.
- Pleural friction rub often present.
- Clubbing of fingers may occur with subacute or chronic lung abscess.
- Gingivodental disease in majority.
- Obstruction of bronchial drainage often caused by stricture, carcinoma or inspissated mucus.
- WBC often elevated to 20,000 to 30,000.
- X-ray shows segmental consolidation, later becomes round infiltrate as area distends with pus; often air fluid level with rupture into bronchial tree.

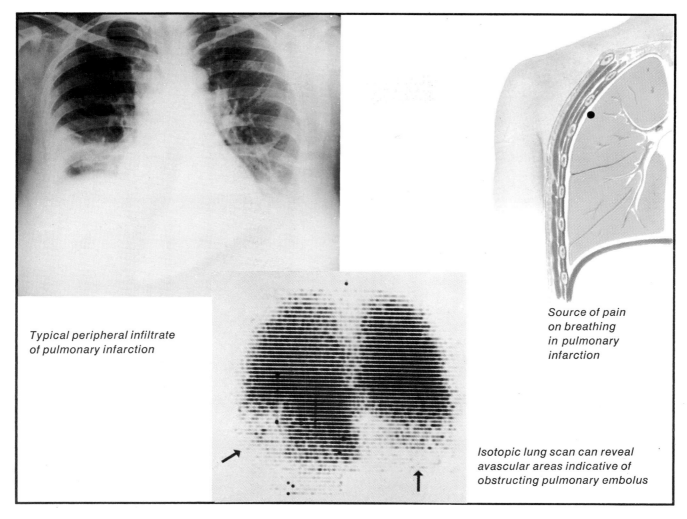

Typical peripheral infiltrate of pulmonary infarction

Source of pain on breathing in pulmonary infarction

Isotopic lung scan can reveal avascular areas indicative of obstructing pulmonary embolus

Chest pain on breathing due to
PULMONARY INFARCTION

Emboli *per se* produce pulmonary infarcts only about 10% of the time. Infarction usually results when the embolus occludes a medium-sized pulmonary artery and collateral circulation is compromised by congestive heart failure or other pulmonary vascular dysfunction. Infarcts are usually preceded by symptoms of embolism — midline chest pain, or merely rapid breathing and tachycardia.

- Pain sharp and jabbing only on breathing; markedly intensified by deep breathing, movement of chest, sneezing or coughing.
- Usually localized to chest wall over infarct; in trapezius muscle and supraclavicular fossa when infarct is adjacent to midportion of diaphragm.
- Onset acute, usually first noticed when patient turns over in bed; persists for hours to days.

ASSOCIATED FINDINGS

- Dyspnea with shallow breathing and nonproductive cough.
- Hemoptysis within 24 hours after onset of pain in 40%; at first bright red, later dark.
- Fever typical but not *always* present.
- Pleural friction rub in about 25%; patch of inspiratory rales more typical.
- Signs of consolidation or pulmonary hypertension in less than 25% of patients.
- Pleural effusion several days after onset of pain; may be serous or bloody on aspiration.
- ECG: normal or transient shift of QRS axis to right, right bundle branch block, atrial fibrillation.
- X-ray may be abnormal in 12 to 24 hours — wedge-shaped or patchy infiltrate; elevated diaphragm; effusion at times.
- Lung scan may show blocked circulation to affected area although x-ray is normal.
- SGOT normal, except with acute right ventricular failure and hepatic congestion.
- Serum bilirubin often rises after 3 or 4 days.
- Serum LDH begins to rise within 24 hours, peaks in 2 or 3 days; rise may persist for 10 days.

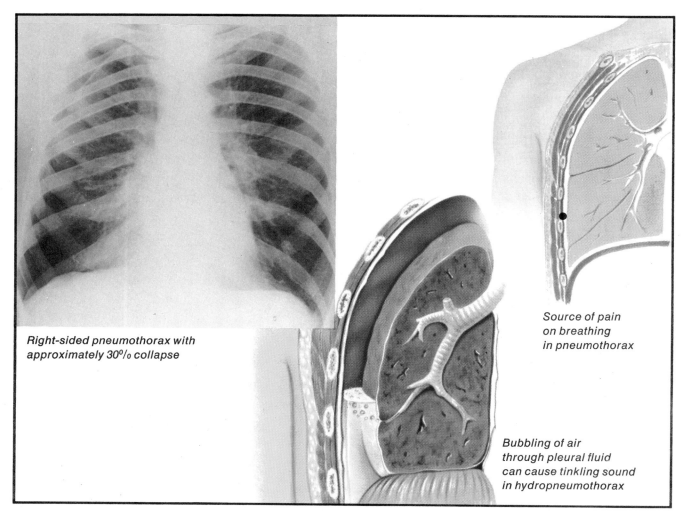

Right-sided pneumothorax with approximately 30% collapse

Source of pain on breathing in pneumothorax

Bubbling of air through pleural fluid can cause tinkling sound in hydropneumothorax

Chest pain on breathing due to
PNEUMOTHORAX

While dyspnea is an important feature of pneumothorax, chest pain on breathing is even more characteristic and more striking. Spontaneous pneumothorax, the most common type, is more prevalent in young men. It may occur following exertion, violent coughing or even at rest.

- Pain sharp and jabbing on breathing; may not disappear when breath is held.
- Intensified by deep breathing, coughing and exertion; not as much as in pneumonia.
- Becomes dull and persistent in later stages, unaffected by respiration.
- Localized in lower part of axilla or under scapula.
- May radiate to base of neck but not down arm.
- Onset acute; persists for hours to days.

ASSOCIATED FINDINGS

- Persistent dyspnea unaffected by position; rapid and shallow breathing with grunting at times; ordinarily no cough; cyanosis in marked cases.
- Chest motion lags on affected side.
- On involved side, breath sounds diminished and amphoric or absent; tactile fremitus decreased; percussion note tympanitic; lower border of percussion does not vary with respiration.
- Large pneumothorax can shift mediastinal contents and cause trachea to deviate to opposite side.
- When extreme, hemithorax is large, interspace wide, percussion note flat; sometimes bulging of intercostal spaces, more apparent in thin persons.
- With stethoscope on posterior chest and coins tapped on anterior chest, ringing note is heard (positive coin sign).
- Metallic tinkle may be heard when air bubbles from bronchopleural fistula come through fluid (hydropneumothorax).
- X-ray shows peripheral lucency of air space; appearance of underlying lung usually normal; small pneumothorax seen only on expiratory x-rays.
- X-ray may be similar to large bulla; but that condition is not acute and coin sign is negative.

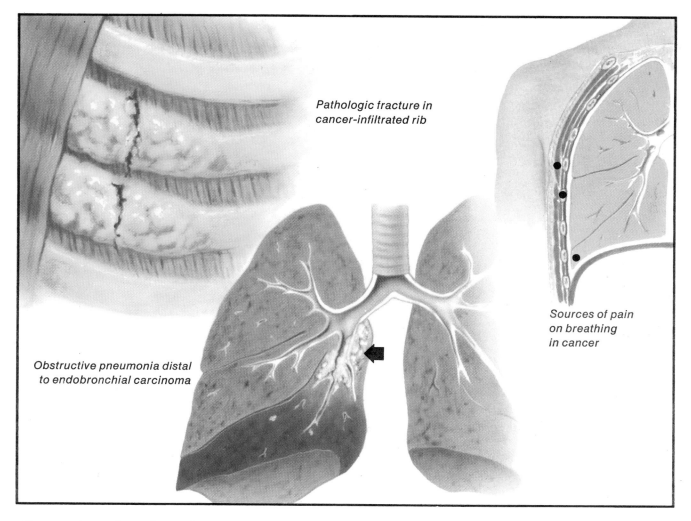

Pathologic fracture in cancer-infiltrated rib

Sources of pain on breathing in cancer

Obstructive pneumonia distal to endobronchial carcinoma

Chest pain on breathing due to
CANCER

Cancer affecting the pleura often produces chest pain on breathing, as a result of an obstructive pleuritic pneumonia or lung abscess, or by direct invasion of the pleura by bronchogenic carcinoma or mesothelioma without symptoms of acute infection. Such pain may also be produced by pathologic rib fractures or by pressure of a tumor mass in the chest wall.

- Pain sharp and jabbing.
- Only on breathing if due to infection; otherwise, may be only on breathing or may be continuous and intensified by breathing, sneezing or coughing.
- Pain localized to affected rib in fracture; dull ache in region of tumor mass, may be localized along course of segmental nerve in chest wall tumor.
- Onset of pain acute with infection or fracture; gradual with tumor in chest wall.

ASSOCIATED FINDINGS

With Pneumonia or Lung Abscess
- Often preceded by chronic cough, hemoptysis and localized wheezing.
- Dyspnea with rales.
- Unlike usual pneumonia — cough often nonproductive, tactile fremitus decreased over involved lobe, breath sounds diminished or absent, resolution slower.
- Pulmonary lesion seen on x-ray.

Pathologic Rib Fracture
- Breathing shallow, proportional to pain.
- Coughing or sneezing often precipitates break.
- Pain reproducible by pressure on fracture or distal area of same rib, but not on adjacent rib.
- Usually no displacement of rib, no crepitation.
- Fracture seen on x-ray.

Tumor Mass in Chest Wall
- Breathing shallow, proportional to pain.
- Palpable mass more likely in rib than in cartilage; may pulsate.
- Lysis of rib and usually soft tissue mass on x-ray.

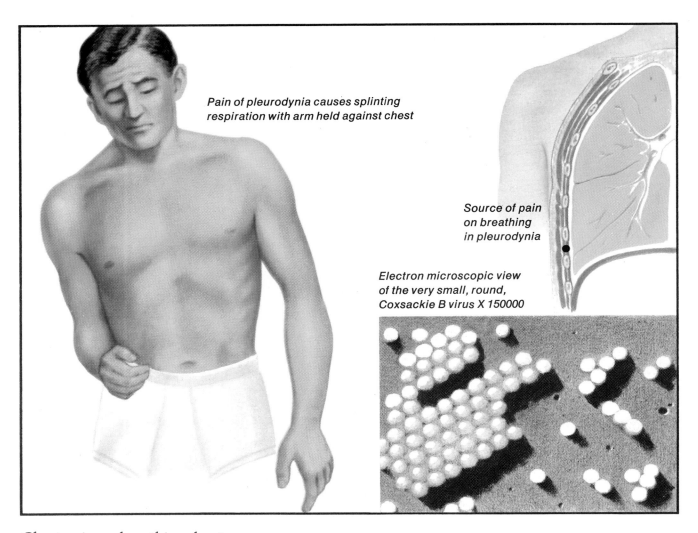

Pain of pleurodynia causes splinting respiration with arm held against chest

Source of pain on breathing in pleurodynia

Electron microscopic view of the very small, round, Coxsackie B virus X 150000

Chest pain on breathing due to
PLEURODYNIA

Pleurodynia, an acute epidemic myalgia or myositis, can be produced by several types of Coxsackie B virus, especially in children and young adults. Also called Bornholm disease or devil's grip, the condition occurs mostly during late summer and early autumn. The typical incubation period is three to five days, but may be as long as two weeks.

- Pain sharp and lancinating only on breathing; intensified by deep breathing and other muscular movements.
- Maximal along attachment of diaphragm; often bilateral, shifting from area to area; frequently radiates to upper abdomen; may simulate acute abdomen in children.
- Onset sudden; episodes usually last for a few hours, recur for 4 to 7 days.
- Young adults lean forward toward painful side to limit motion of chest, often holding arm against painful area; children often draw up knees.

ASSOCIATED FINDINGS

- Muscular aches often present in neck, shoulders, back and gluteals.
- Breathing rapid and shallow.
- Pharyngitis, fever, nausea and vomiting in some.
- No deep tenderness over painful area, but superficial hyperesthesia may be present.
- Pleuritis manifested by pleural friction rub in about 25% of patients.
- Occasional nuchal rigidity and headache.
- Meningitis may occur several days after onset of fever.
- Orchitis an occasional complication.
- Myocarditis and pericarditis (caused by same virus) rarely accompany pleurodynia.
- WBC varies from normal to 20,000; often eosinophilia.
- Chest x-ray and ECG usually normal.
- Virus can be recovered from throat washings, stool, blood, urine and cerebrospinal fluid; definitive diagnosis by rising titer of neutralizing or complement-fixing antibodies in serum.

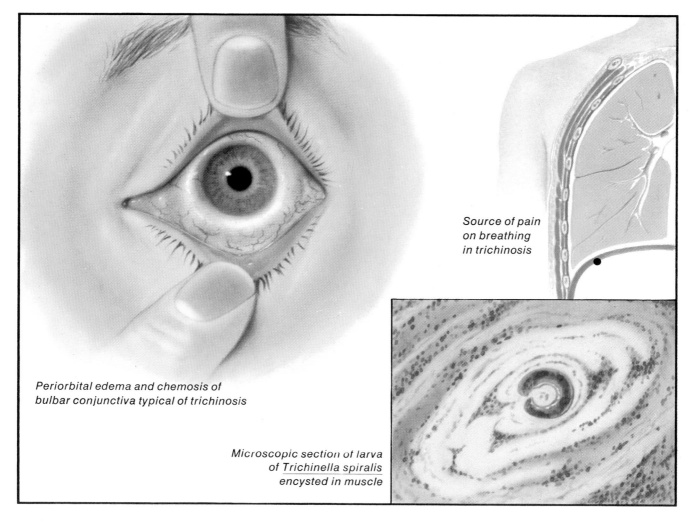

Periorbital edema and chemosis of bulbar conjunctiva typical of trichinosis

Source of pain on breathing in trichinosis

Microscopic section of larva of *Trichinella spiralis* encysted in muscle

Chest pain on breathing due to
TRICHINOSIS

Migration and invasion of larvae of the roundworm, *Trichinella spiralis*, produces muscular inflammation and pain. This may follow ingestion of inadequately cooked infected pork by seven to nine days. The skeletal, gluteal, pectoralis, deltoid, gastrocnemius, intercostal, laryngeal, lingual or masseter muscles can be involved. Chest pain on breathing occurs when the diaphragm is involved.

- Pain severe and sharp only on breathing or movement of chest.
- Primarily localized to chest wall insertion of diaphragm; usually bilateral.
- Onset gradual; becomes progressively more severe.
- Severe pain may last two months, followed by dull aching pain for months to years.

ASSOCIATED FINDINGS

- Breathing shallow, proportional to pain; dyspnea may also result from pulmonary edema.
- Stage of invasion — nausea, vomiting, diarrhea and fever usually a week prior to pain.
- Stage of migration — muscular pain with marked fever; nausea, vomiting and diarrhea at times.
- Headache, delirium and psychosis common.
- Generalized edema may be observed.
- Periorbital edema, chemosis of bulbar conjunctivae, and visual disturbances often present.
- Movement of eyes or limbs may be painful.
- Affected muscles tender to pressure.
- PR prolongation and ST-T abnormalities on electrocardiogram.
- WBC normal or elevated; marked eosinophilia after 14 days.
- X-ray shows calcification of cysts in muscles after six months.
- Definitive diagnosis by muscle biopsy.
- Symptoms may resemble influenza or encephalitis.

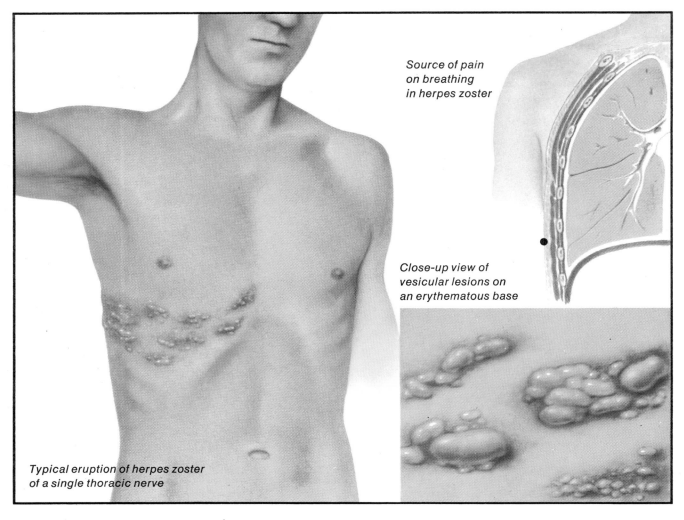

Source of pain on breathing in herpes zoster

Close-up view of vesicular lesions on an erythematous base

Typical eruption of herpes zoster of a single thoracic nerve

Chest pain on breathing due to
HERPES ZOSTER

Herpes zoster, an inflammation caused by the varicella virus, is usually limited to a single nerve and the skin distribution of that nerve. It can involve either thoracic or abdominal spinal nerves. With thoracic involvement, there is chest pain on breathing. Herpes zoster usually occurs after the age of 20, at times as a complication of arsenic poisoning, tuberculosis, cancer or lymphoma. In many patients there may be no apparent predisposing factor. It is most common in late winter and early spring.

- Pain severe, burning and stabbing.
- May be continuous and intensified by breathing.
- Localized along a thoracic nerve distribution on one side, with hyperesthesia.
- Pain usually precedes eruption by 3 or 4 days; sometimes abates when rash appears.
- Pain and lesions last for several weeks, usually disappear together.
- Neuralgic and often intractable postherpetic pain may persist for years, especially in the aged.

ASSOCIATED FINDINGS

- Breathing shallow, proportional to pain.
- Movement of chest limited.
- Slight fever and malaise often present.
- Regional nodes may enlarge before rash appears.
- Tenderness along involved nerve before, during and after rash.
- Vesicular eruptions, singly or in confluent groups, on erythematous base along involved nerve.
- Vesicles break and form scabs.
- Scarring often follows.
- Stained scrapings of young vesicles show giant cells — multinucleate with intranuclear inclusions.
- Severe involvement can affect motor roots and produce temporary or permanent paralysis.

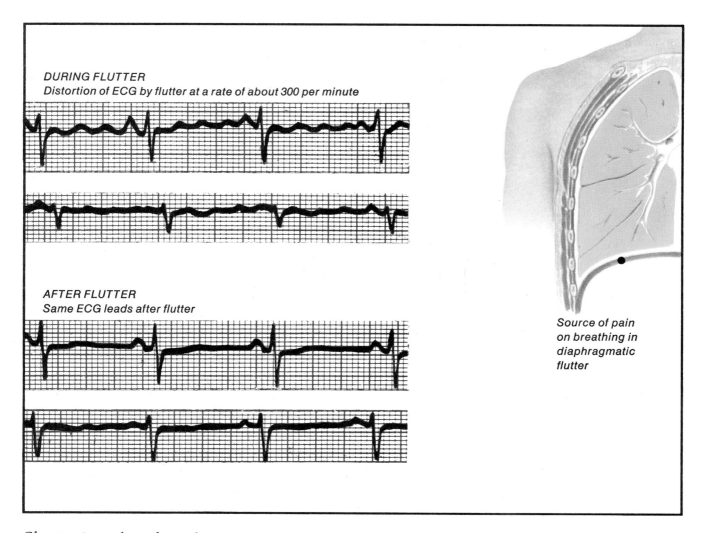

DURING FLUTTER
Distortion of ECG by flutter at a rate of about 300 per minute

AFTER FLUTTER
Same ECG leads after flutter

Source of pain on breathing in diaphragmatic flutter

Chest pain on breathing due to
DIAPHRAGMATIC FLUTTER

Episodes of rapid, regular or irregular contractions of the diaphragm — continuously or in paroxysms — can occur following encephalitis, fractures of the lower rib cage or xiphoid, or hyperventilation tetany. They may be psychogenic in origin. In some cases, the flutter is precipitated by deep breathing, coughing, sneezing, emotional disturbances, pressure on the epigastrium or exercise. The flutter is sometimes suppressed by deep inspiration, swallowing, supraclavicular pressure or sleep.

- Pain severe and sharp.
- May be continuous and intensified by breathing, or may be only on breathing.
- Along one or both sides of diaphragmatic attachment.
- Occasionally midline and pressing, simulating cardiac ischemic pain.
- May radiate to epigastrium, neck, shoulders and arms.
- Onset of episode sudden.
- Duration variable, minutes to hours.

ASSOCIATED FINDINGS

- Breathing shallow, proportional to pain.
- Dyspnea when *slow* flutter rate controls respiration; respiratory alkalosis may follow.
- Flutter rate may vary from 35 to 480 per minute.
- Excursion may vary from a few millimeters to several centimeters.
- Flutter may involve whole or part of diaphragm; usually superimposed on normal motion.
- A "to and fro" sound, simulating a secondary heart beat, can be heard at times.
- Sound of rapid splashing of stomach contents in some patients.
- Flutter best seen on fluoroscopy.
- ECG often normal; sometimes nonspecific ST-T changes or deflections synchronous with motion of diaphragm.

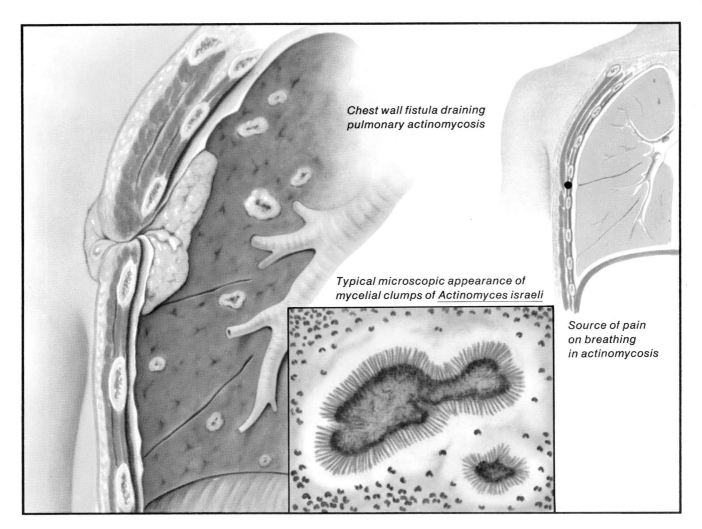

Chest wall fistula draining pulmonary actinomycosis

Typical microscopic appearance of mycelial clumps of Actinomyces israeli

Source of pain on breathing in actinomycosis

Chest pain on breathing due to ACTINOMYCOSIS

A chronic granulomatous disease of infectious origin, actinomycosis affects the cervicofacial, abdominal or thoracic regions; it is characterized by abscess formation with multiple draining sinuses. *Actinomyces israeli* are usually found in involved tissues and exudates. Thoracic involvement is either a direct extension of the abdominal infection or a result of aspiration of infected material from a cervicofacial lesion. The pleura and the thoracic wall are generally involved, the mediastinum rarely. Chest pain on breathing occurs with pleuritic involvement.

- Pain sharp and pleuritic; may be continuous and intensified by breathing and movement of chest or may occur only on breathing.
- Localized to involved area; does not radiate.
- Onset gradual; usually does not become severe.
- Prolonged in duration; may be recurrent.

ASSOCIATED FINDINGS

- Breathing shallow, proportional to pain.
- Physical signs suggestive of subacute pulmonary infection.
- Recurrent high fever and night sweats.
- Progressive wasting frequently present.
- Chronic productive cough; mucopurulent sputum, at times fetid.
- Dullness to percussion and inspiratory crepitant rales.
- Pleural effusion and empyema rare.
- Secondary bacterial infections common.
- Destruction of adjacent rib and draining fistulas in chest wall may develop.
- Actinomycotic lesions may appear on skin and soft tissues of face and neck, in the cecum or appendix.
- Pus from fistula contains mycelial clumps, "sulfur granules"; microscopic examination reveals gram-positive branching hyphae, often beaded.
- Definitive diagnosis by culture of causative agent.

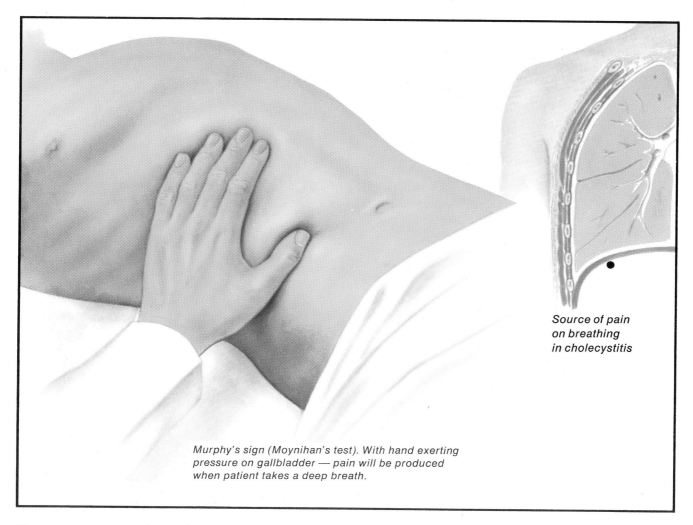

Murphy's sign (Moynihan's test). With hand exerting pressure on gallbladder — pain will be produced when patient takes a deep breath.

Source of pain on breathing in cholecystitis

Chest pain on breathing due to
CHOLECYSTITIS

Chest pain on breathing usually occurs with acute inflammation of the gall bladder, whether or not associated with stones. Pleuritic chest pain is much less likely in biliary colic, which is characterized by wavelike lateral chest and epigastric pain. Both conditions are common in women between 30 and 50 years of age.

- Pain sharp and jabbing; may be continuous and intensified by breathing or may occur only on breathing.
- Usually localized in right lower chest.
- Maximal in right upper quadrant of abdomen.
- May radiate to right axilla and shoulder.
- Gradual in onset; sudden when stones are present.
- Persistent; may last several days (colic usually lasts only 3 to 12 hours).

ASSOCIATED FINDINGS

- Breathing shallow, proportional to pain.
- Fear of exacerbation of pain restricts movement, particularly of right side.
- High fever.
- Nausea and vomiting.
- Constipation frequent.
- Jaundice if stones are present.
- Right upper quadrant of abdomen tender to palpation.
- Muscle spasm and even rebound tenderness in severe attacks.
- Abdominal distention and diminished bowel sounds.
- Enlarged gallbladder may be palpable, especially on first attack.
- Upon deep breath, while gallbladder is pressed, patient feels painful "catch" as diaphragm descends.
- Urine urobilinogen may be elevated, even when jaundice is not present.
- WBC elevated, up to 40,000, with many band forms.

DIFFERENTIAL DIAGNOSIS OF CHEST PAIN ON BREATHING

	Character	Location	Radiation	Onset
ACUTE PLEURISY	Sharp and jabbing only on breathing	Lower lateral chest	Back, shoulders, arms, occasionally to abdomen	Acute
PNEUMONIA	Sharp and jabbing only on breathing	Over involved lobe; usually lower lateral chest	Usually none	Acute
LUNG ABSCESS	Usually sharp and jabbing only on breathing; at times dull and aching	Over abscess	None	Gradual
PULMONARY INFARCTION	Sharp and jabbing only on breathing	Over infarct; in trapezius muscle and supraclavicular fossa with central infarcts	None	Acute
PNEUMOTHORAX	Sharp and jabbing; may be continuous; intensified by breathing	Lower axilla or under scapula	Base of neck	Acute
CANCER	Sharp and jabbing only on breathing; also dull and aching with tumor mass	Over lesion or course of involved segmental nerve	None	Acute; gradual with tumor mass

Duration	Breathing	Cough	Associated Signs and Symptoms	Key Laboratory Data
Usually few days	Shallow and rapid (proportional to pain)	Nonproductive	Fever; pleural friction rub; signs of effusion; usually associated with tuberculosis, collagen disease, carcinoma; may occur with or without effusion	Tuberculin test positive in tuberculous pleurisy; effusion high in LDH in subpleural bronchogenic carcinoma
Several days	Severely dyspneic	Severe and hacking; sputum purulent	Fever; pleural friction rub; rales; signs of consolidation; pain prominent in bacterial pneumonia; not a feature of diffuse viral pneumonia	Elevated WBC; increased polymorphonuclear leukocytes; x-ray shows opacification of consolidated lobe; specific diagnosis by smear and culture of sputum
Several days	Shallow (proportional to pain)	Copious sputum, purulent, often fetid and/or bloody	Fever; pleural friction rub; rales; dullness to percussion; may follow aspiration; most cases bronchogenic in origin; 25% due to bronchostenosis of bronchogenic carcinoma	WBC 20,000 to 30,000; x-ray: segmental consolidation, becomes round infiltrate, often air fluid level if rupture occurs
Hours to days	Dyspneic	Nonproductive or with hemoptysis	Fever; pleural friction rub; rales; signs of effusion	SGOT usually normal; serum bilirubin and LDH rise; wedge-shaped or patchy infiltrates on x-ray; ECG shows transient shift of QRS axis to right
Hours to days	Persistently dyspneic	Ordinarily absent	Diminished breath sounds; tympanitic or flat to percussion; tactile fremitus decreased	X-ray shows peripheral lucency of airspace; coin test positive
Few days; prolonged with tumor mass	Shallow (proportional to pain); dyspneic with obstructive pneumonia	Frequent with bronchial lesion	Pain of rib fracture reproducible; tumor mass palpable; rales with obstructive pneumonia and abscess	Fracture, lytic lesion, tumor mass or pulmonary lesion seen on x-ray

(Continued on next page.)

DIFFERENTIAL DIAGNOSIS OF CHEST PAIN ON BREATHING (continued)

	Character	Location	Radiation	Onset
PLEURODYNIA	Sharp and lancinating only on breathing	Along attachment of diaphragm; often bilateral	Upper abdomen	Sudden
TRICHINOSIS	Severe and sharp only on breathing	Along chest wall insertion of diaphragm; usually bilateral	None	Gradual
HERPES ZOSTER	Severe, burning and stabbing; may be continuous; intensified by breathing	Along thoracic nerve distribution on one side	None	Acute; usually precedes eruption
DIAPHRAGMATIC FLUTTER	Severe and sharp; may be continuous; intensified by breathing	Along one or both sides of diaphragmatic attachment; at times midline	Epigastrium, neck, shoulders, arms	Sudden
ACTINOMYCOSIS	Sharp and pleuritic; may be continuous; intensified by breathing	Over involved area	None	Gradual
CHOLECYSTITIS	Sharp and jabbing; may be continuous; intensified by breathing	Right lower chest or right upper quadrant of abdomen	None; biliary colic radiates to right axilla and shoulders	Gradual; sudden when stones are present

Duration	Breathing	Cough	Associated Signs and Symptoms	Key Laboratory Data
Episodes of few hours; recur for 4-7 days	Rapid and shallow (proportional to pain)	None	Fever; pleural friction rub	WBC may be elevated; chest x-ray and ECG normal
Months	Shallow (proportional to pain)	Usually none	Fever; generalized and facial edema; affected muscles tender; may simulate influenza or encephalitis	Skin and serological tests positive; marked eosinophilia; x-ray shows calcification of cysts in muscles after six months; muscle biopsy for definitive diagnosis
Several weeks	Shallow (proportional to pain)	None	Slight fever and malaise; vesicular erythematous eruptions; enlarged regional nodes; tenderness along involved nerve	Giant cells — multinucleate with intranuclear inclusions on stain of vesicle scraping
Minutes to hours	Shallow (proportional to pain); dyspneic when flutter controls respiration	None	"To and fro" sound simulating a secondary heart beat at times; sound of rapid splashing of stomach contents in some	Flutter best seen on fluoroscopy; ECG may show deflections synchronous with motion of diaphragm
Prolonged; recurrent	Shallow (proportional to pain)	Chronic with mucopurulent sputum, at times fetid	Fever; dullness to percussion; rales; draining fistula may be present; actinomycotic lesions on face or neck	Gram-positive hyphae in stained "sulfur granules" from pus of fistula
Several days	Shallow (proportional to pain)	None	Fever; nausea; vomiting; constipation; jaundice if stones are present; right upper quadrant of abdomen tender to palpation	WBC elevated with many band forms; urine urobilinogen elevated

COUGH

Cough serves the useful function of expelling foreign material or excessive secretions from the respiratory tract. It is a reflex only partly under voluntary control, which can be stimulated anywhere along the respiratory mucous membranes or from the pleura, esophagus, pharynx or even the ear.

Most coughs result from relatively minor conditions of the pharynx and larynx; many are psychogenic in origin. However, cough can be caused by obstruction, compression, inflammation or infection in the lower respiratory tract, and can be a symptom of a serious illness. Characterization of the nature of the cough and analysis of associated findings can facilitate correct diagnosis.

Twelve important causes of coughing are described individually and then compared in tabular form.

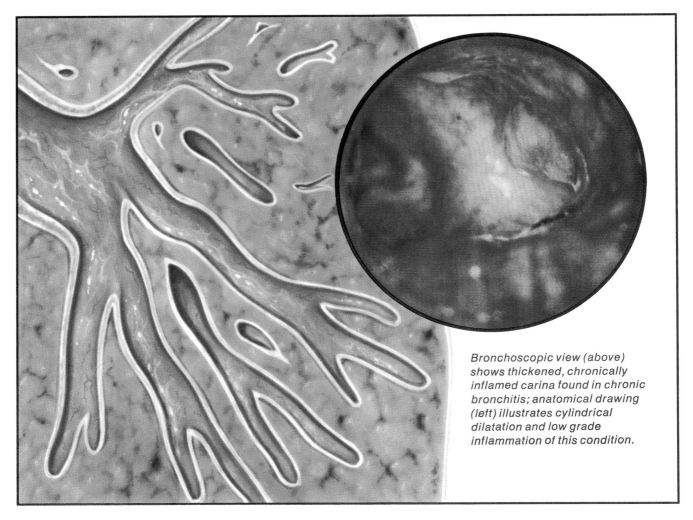

Bronchoscopic view (above) shows thickened, chronically inflamed carina found in chronic bronchitis; anatomical drawing (left) illustrates cylindrical dilatation and low grade inflammation of this condition.

Cough due to
CHRONIC BRONCHITIS

A persistent productive cough, present for at least three months of the year for two or more consecutive years, characterizes this condition. Chronic bronchitis is commonly associated with pulmonary emphysema. It usually occurs in older people (after forty), and is four times more frequent in men than in women. Cigarette smoking, air pollution and cold damp weather are important predisposing factors.

- Onset of cough insidious; may be initial symptom.
- At first, intermittent "chest cough" following URI; becomes severe, at times paroxysmal.
- Variable — mild in summer, severe in winter; aggravated by URI.
- Usually productive.
- Sputum tenacious, mucoid or mucopurulent.
- Hemoptysis uncommon.
- Cough more marked during night and on arising.
- Chronic and persistent.

ASSOCIATED FINDINGS

- Fever absent; malaise slight.
- Progressive exertional dyspnea.
- Breath sounds roughened; transient expiratory wheezes heard only on forced expiration at times.
- Few inspiratory fine or medium moist rales; if numerous or chronic, bronchiectasis should be suspected.
- No signs of consolidation.
- Paranasal sinusitis and posterior rhinitis often associated.
- Chest x-ray may be normal or show increased hilar markings; may reveal hyperlucency and depressed diaphragm in presence of emphysema.
- Bronchogram will show cylindrical bronchial dilatation, not saccular as in bronchiectasis.
- Bronchoscopy will reveal normal or slightly reddened bronchial mucosa.

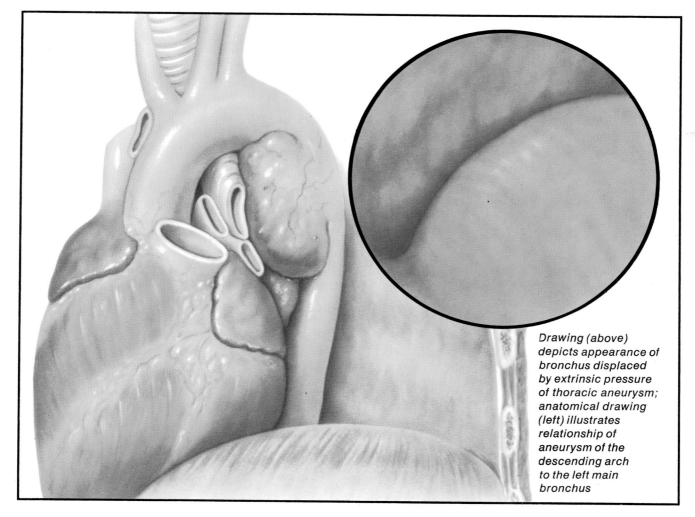

Drawing (above) depicts appearance of bronchus displaced by extrinsic pressure of thoracic aneurysm; anatomical drawing (left) illustrates relationship of aneurysm of the descending arch to the left main bronchus

Cough due to
THORACIC AORTIC ANEURYSM

Cough is the earliest and most consistent symptom of thoracic aortic aneurysm. It is produced by pressure of the aneurysm on the trachea, bronchi or lung parenchyma. Syphilis is the most likely cause in patients over age 40 (particularly males). In younger patients Marfan's syndrome must be suspected.

- Onset insidious.
- "Brassy" when trachea is compressed.
- Not severe or paroxysmal; usually nonproductive.
- When productive (due to compressed bronchi with obstructive infection), sputum is mucopurulent.
- Cough unaffected by position or time of day; chronic and persistent.

ASSOCIATED FINDINGS

- Fever absent.
- Anterior chest pain with sternal bone erosion; lateral chest pain with rib erosion; girdling neuritic pain with intercostal nerve compression.
- Pain in right shoulder and upper extremity will suggest extension to innominate artery.
- Dysphagia when esophagus compressed; inspiratory stridor when trachea compressed; chronic hiccup when phrenic nerve involved; hoarseness with left vocal cord paresis.
- Percussion of aneurysm usually difficult.
- Basal diastolic murmur of aortic insufficiency often.
- "Tracheal tug" with each heart beat, when trachea is compressed.
- Plethora and distended veins in head, neck, arms and chest, when superior vena cava compressed.
- Right heart failure, when pulmonary artery compressed.
- X-ray shows fusiform or saccular dilatation of aorta; often linear calcification of ascending aorta.
- Serologic test for reagin usually positive; specific serologic test (TPI, FTA) nearly always positive when aneurysm is due to syphilis.
- Great height, long hyperextensible fingers, high-arched palate and prolapsed opacified lenses may occur with Marfan's syndrome.

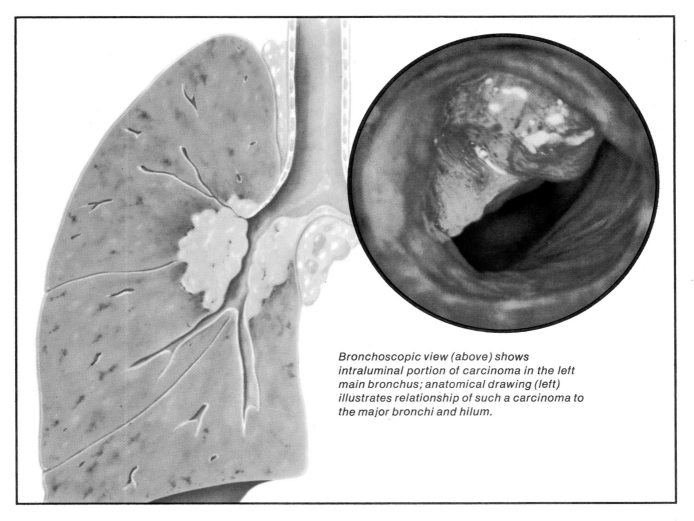

Bronchoscopic view (above) shows intraluminal portion of carcinoma in the left main bronchus; anatomical drawing (left) illustrates relationship of such a carcinoma to the major bronchi and hilum.

Cough due to
BRONCHOGENIC CARCINOMA

Cough is an important presenting symptom in many cases of bronchogenic carcinoma. Its diagnostic significance can be overlooked in a patient with chronic bronchitis. A preexisting cough usually changes in character and severity because of irritation or obstruction by the endobronchial lesion. Bronchogenic carcinoma usually occurs in middle-aged and older men.

- Onset insidious.
- Frequent, but usually not paroxysmal.
- May be mild at first, later may become severe; may be nonproductive, usually becomes productive.
- Sputum mucoid or mucopurulent, may be bloody; hemoptysis common, may be initial symptom.
- Cough unaffected by position or time of day.
- Chronic and persistent.

ASSOCIATED FINDINGS

- Fever absent except with associated obstructive pneumonia or lung abscess.
- Weakness and malaise may be presenting symptoms.
- Chest pain often present (pleuritic or nondescript).
- Weight loss, cachexia, anemia in latter stages.
- Recurrent pneumonias, lung abscess, bronchiectasis may occur distal to tumor.
- Localized reduction in breath sounds, late onset of inspiratory sound and wheezing.
- Hoarseness, chest lag, enlarged supraclavicular nodes at times.
- Clubbing of fingers, pulmonary osteoarthropathy, peripheral neuropathy, thrombophlebitis at times.
- Plethora and distended veins in neck, face, arms and chest — with occlusion of superior vena cava.
- Atrophy of muscles of arm and hand, neuritic pain and Horner's syndrome — with tumor in lung apex.
- X-ray shows new density without calcification, hilar accentuation, mediastinal widening or pleural effusion.
- Can be diagnosed by bronchoscopy in 50% of cases.
- Tumor cells often identifiable in sputum.

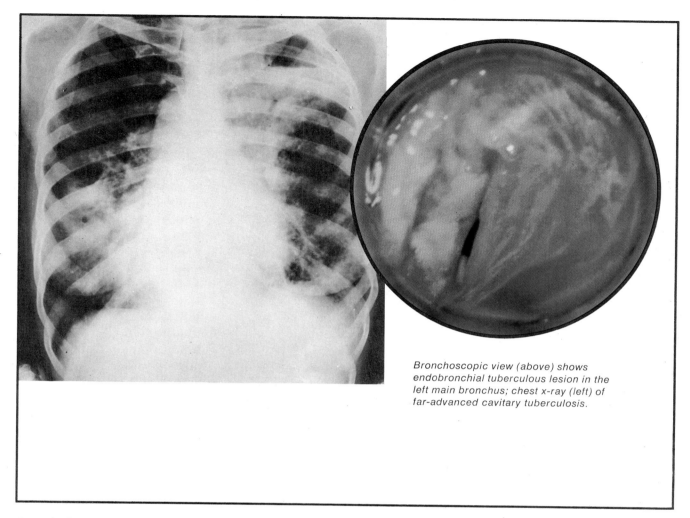

Bronchoscopic view (above) shows endobronchial tuberculous lesion in the left main bronchus; chest x-ray (left) of far-advanced cavitary tuberculosis.

Cough due to
PULMONARY TUBERCULOSIS

A chronic productive cough is a prominent symptom of moderate and advanced adult-type pulmonary tuberculosis. In the presence of mild fever and malaise, this condition can often be misdiagnosed as chronic bronchitis. Cough is not a feature of primary tuberculosis in children or of miliary tuberculosis, unless large hilar nodes compress or erode the bronchi.

- Onset insidious.
- Initially dry and hacking; usually becomes severe and productive.
- Frequent but usually not paroxysmal.
- Sputum mucopurulent; often green.
- Hemoptysis common; massive hemorrhage occasional.
- Cough usually not affected by position or time of day; chronic and persistent.

ASSOCIATED FINDINGS

- Fever and night sweats common; fever can be high without acute toxicity.
- Malaise, weakness, easy fatigability, weight loss common.
- Suppression of menses frequent.
- Fine moist rales, at times audible only after cough, in early stages; dullness to percussion and inspiratory fine and medium moist rales with extensive involvement.
- Narrowing of band of supraclavicular percussive resonance with apical disease.
- Cavernous breath sounds and tympanitic percussion note over cavity when cavity walls are rigid near periphery and communicate with bronchus.
- "Cracked-pot" resonance by percussing over cavity and listening with stethoscope over open mouth.
- Smear and culture of sputum, or culture of gastric contents, are diagnostic.
- X-ray shows patchy infiltrates, at times with areas of fibrosis or cavitation, and compensatory emphysema.
- Tuberculin test almost always positive in adult-type pulmonary tuberculosis.

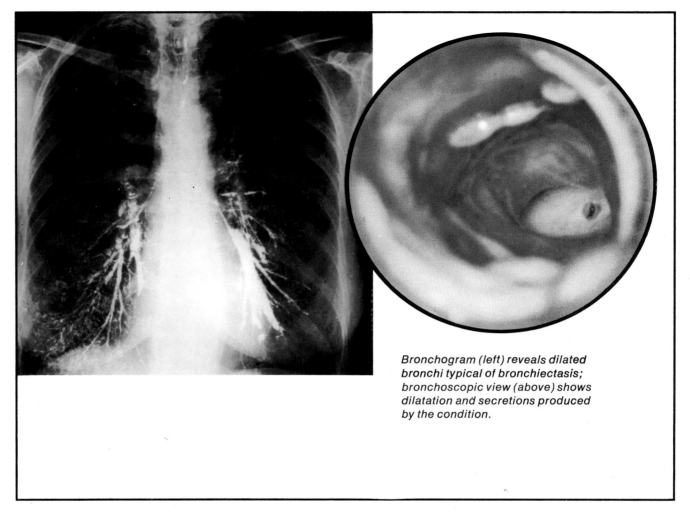

Bronchogram (left) reveals dilated bronchi typical of bronchiectasis; bronchoscopic view (above) shows dilatation and secretions produced by the condition.

Cough due to
BRONCHIECTASIS

A chronic productive cough, maximal on arising, is characteristic of bronchiectasis. This disorder should be suspected when a chronic productive cough is associated with recurrent pneumonias and hemoptysis. Usually following necrotizing pneumonia or infection of congenital cysts, it may be a complication of tuberculosis or bronchogenic carcinoma. It differs from bronchitis in that there is saccular dilatation of the bronchi without generalized airway obstruction.

- Onset insidious.
- Prolonged paroxysms (minutes to hours).
- Variable — may be mild, moderate or severe.
- Productive; may seem nonproductive when sputum is swallowed; nonproductive in "dry" bronchiectasis; sputum copious and purulent; often fetid and/or bloody; less tenacious than in lung abscess.
- Hemoptysis common; massive hemorrhage rare.
- Cough usually on arising, lying or bending.
- Chronic and persistent.

ASSOCIATED FINDINGS

- Fever absent, except with superimposed pneumonia.
- Dyspnea and cyanosis absent, except with bronchitis and emphysema.
- Clubbing of fingers occasionally seen.
- Medium or coarse moist rales and rhonchi; occasionally amphoric breath sounds over large cystic dilatations.
- Intermittent episodes of pneumonia in bronchiectatic area, with signs of consolidation, fever, pleuritic pain.
- Sputum "layers out" on standing (mucus on top, pus below).
- Pus cells are polys, not eosinophils.
- Paranasal sinusitis frequent.
- Anemia sometimes present.
- Culture may show various bacteria.
- X-ray will show increased peribronchial markings in basilar segments of lower lobe; saccular dilatations on bronchogram are diagnostic.

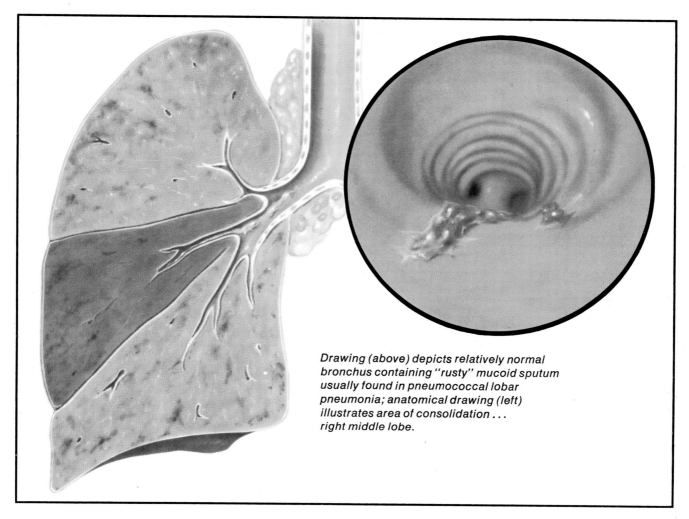

Drawing (above) depicts relatively normal bronchus containing "rusty" mucoid sputum usually found in pneumococcal lobar pneumonia; anatomical drawing (left) illustrates area of consolidation... right middle lobe.

Cough due to PNEUMOCOCCAL LOBAR PNEUMONIA

Cough is usually present from the onset of pneumococcal lobar pneumonia. It may, however, be overshadowed by other symptoms in the initial stages. Pneumococcal lobar pneumonia is usually preceded by an upper respiratory infection. Its onset is abrupt with rigor, chills, fever, pleuritic chest pain and prostration.

- Onset of cough usually insidious.
- Paroxysmal at times.
- Initially slight, becomes severe.
- Gradually becomes productive.
- Sputum tenacious, mucoid and "rusty"; becomes loose and mucopurulent, often yellow.
- Cough unaffected by position or time of day.
- Persists for 8 to 10 days if untreated.

ASSOCIATED FINDINGS

- Marked fever and prostration; often follows severe chill.
- Headache, even delirium, in severely ill.
- Skin warm and flushed.
- Severe pleuritic pain over involved lobe with tenderness and hyperesthesia of chest wall.
- Severe dyspnea with rapid, shallow breathing and expiratory grunting; cyanosis at times.
- Slight change in whispered voice may be only auscultatory finding in early stages; later, pleural friction rub and signs of consolidation.
- Fine and medium inspiratory moist rales; become coarse and numerous during resolution.
- Signs of fluid not uncommon, empyema in 5%.
- Herpetic blisters around mouth in some.
- Abdominal distention frequent.
- WBC elevated; increase in polys.
- Sputum smear will show encapsulated gram-positive cocci; pneumococci often cultured from both sputum and blood.
- X-ray shows opacification of consolidated lobe.

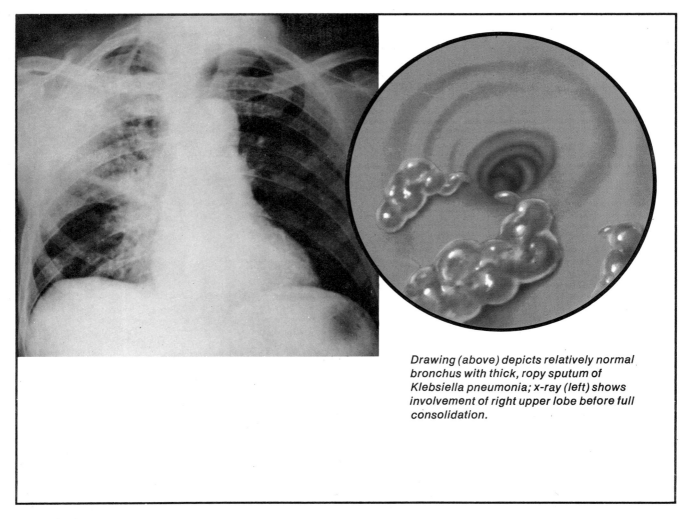

Drawing (above) depicts relatively normal bronchus with thick, ropy sputum of Klebsiella pneumonia; x-ray (left) shows involvement of right upper lobe before full consolidation.

Cough due to
KLEBSIELLA PNEUMONIA

Severe cough, productive of bloody sputum, is a prominent symptom of lobar or confluent pneumonia due to *Klebsiella pneumoniae*. As in pneumococcal lobar pneumonia, the onset is abrupt with rigor, chills, fever, pleuritic chest pain and prostration. Unlike pneumococcal lobar pneumonia, one or more lobes (usually upper) are involved, and it is not preceded by upper respiratory infection. It occurs mostly in alcoholic, debilitated and malnourished patients.

- Onset of cough usually acute.
- Paroxysmal, severe from onset.
- Always productive; sputum copious, tenacious and mucoid; often brick-red (bloody); at times brown and ropy.
- Hemoptysis common; at times marked.
- Cough unaffected by position or time of day; persists throughout course of disease.

ASSOCIATED FINDINGS

- Fever not as high or persistent as in pneumococcal or mycoplasma pneumonia.
- Chill and marked prostration at onset.
- Pleuritic chest pain over involved lobe.
- Dyspnea, rapid shallow breathing and cyanosis occur early and are severe and persistent.
- Jaundice frequent; abdominal distention, vomiting, diarrhea may simulate associated gastrointestinal disease.
- Pleuritic friction rub usually present.
- Signs of consolidation, moist rales, diminished breath sounds — when bronchus partially plugged.
- Shift of trachea toward involved side, marked dullness, absent breath sounds, no rales — when bronchus completely plugged.
- Necrosis and multiple abscesses may occur; cavernous breath sounds absent.
- X-ray shows opacification, often with cavities, of one or more lobes; heavy exudate in right upper lobe may cause sagging short fissure.
- Diagnosis by smear and culture of sputum.

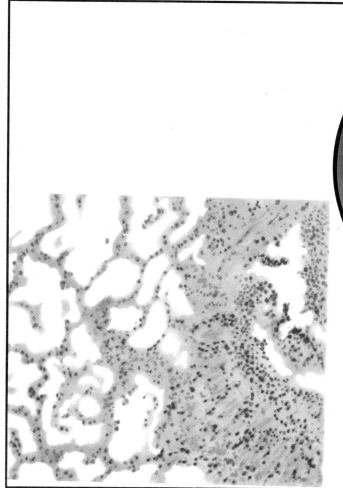
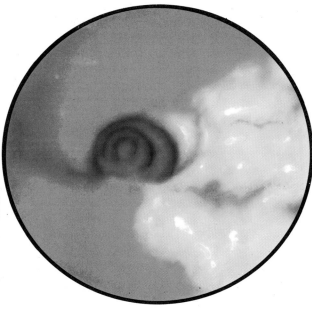

Drawing (above) depicts reddened bronchial mucosa and mucoid sputum usually present in mycoplasma pneumonia; drawing of microscopic section (left) pictures round cell pulmonary and peribronchial inflammation characteristic of condition.

Cough due to
MYCOPLASMA PNEUMONIA

Cough is the most consistent symptom of mycoplasma (primary atypical) pneumonia; its absence makes the diagnosis questionable. This pneumonia occurs sporadically, usually preceded by an upper respiratory infection. It is most common in school-age children and young adults. Older individuals appear to develop immunity. Chest x-ray is most important for correct diagnosis.

- Onset of cough is acute.
- Paroxysmal much of the time.
- At first nonproductive, becomes productive.
- Sputum mucoid or mucopurulent, often blood streaked.
- Frank hemoptysis does not occur.
- Cough unaffected by position or time of day.
- Persists throughout course of disease.

ASSOCIATED FINDINGS

- Marked fever and chilly sensations.
- Headache and malaise may be severe, but no delirium or prostration.
- Substernal rather than pleuritic chest pain.
- Pain aggravated by coughing.
- Dyspnea and cyanosis rare.
- Throat usually sore, nasopharyngeal membrane inflamed with scanty exudate.
- Patchy fine and medium rales may be only auscultatory finding; pleural friction rub or effusion rare.
- Relatively slow heart rate.
- Little if any elevation of WBC; ESR always elevated.
- X-ray shows scattered mottled densities mostly in region of hila and lower lobes, may migrate; x-ray findings more marked than expected from physical exam.
- Cold agglutinins in 50% of cases; special procedures needed to distinguish mycoplasma from viral and Q fever pneumonias.

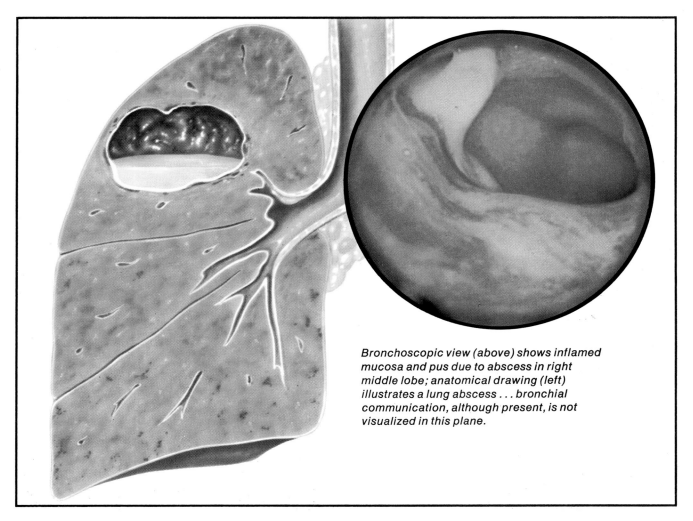

Bronchoscopic view (above) shows inflamed mucosa and pus due to abscess in right middle lobe; anatomical drawing (left) illustrates a lung abscess... bronchial communication, although present, is not visualized in this plane.

Cough due to
LUNG ABSCESS

Cough and pleuritic chest pain are the dominant symptoms of lung abscess. Although the cough, malaise and fever of this condition may mimic acute bronchitis or pneumonia, pleuritic chest pain separates it from bronchitis; and the character of sputum and x-ray appearance help establish the correct diagnosis.

- Onset acute or insidious.
- Cough frequent but usually not paroxysmal.
- Becomes severe.
- At first nonproductive, becomes productive within days; sputum copious, tenacious and purulent; often fetid and/or bloody.
- Hemoptysis common; massive hemorrhage rare.
- Cough usually unaffected by position or time of day; persists for weeks if untreated.

ASSOCIATED FINDINGS

- Fever and malaise always present.
- Sharp and jabbing pleuritic chest pain on breathing for several days during acute stage.
- Breathing shallow, proportional to pain.
- Distinctive physical findings scanty.
- Pleural friction rub often present.
- Dullness to percussion and inspiratory moist rales due to consolidation.
- Amphoric breath or voice sounds not usually present because cavity is thin walled.
- Clubbing of fingers may occur with subacute or chronic lung abscess; usually develops late, 1 to 2 weeks at the earliest.
- Various organisms (including spirochetal and fusiform) may be cultured.
- Gingivodental disease with foul breath frequent.
- Obstruction of bronchial drainage often present from stricture, carcinoma or inspissated mucus.
- X-ray shows round or segmental infiltrate; often, but not always, with air fluid level.
- WBC often elevated — 20,000 to 30,000.

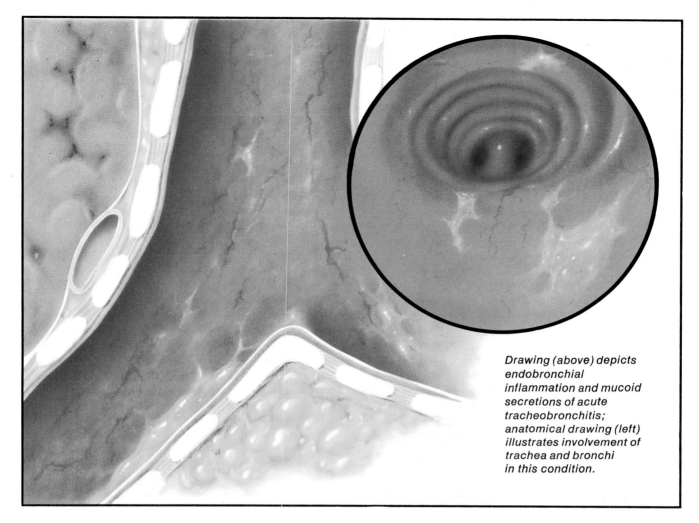

Drawing (above) depicts endobronchial inflammation and mucoid secretions of acute tracheobronchitis; anatomical drawing (left) illustrates involvement of trachea and bronchi in this condition.

Cough due to
ACUTE TRACHEOBRONCHITIS

Cough is a consistent feature of all forms of acute tracheobronchitis. Often the first symptom in pertussis and measles, it usually follows systemic symptoms in influenza.

- Onset of cough insidious in pertussis; acute in influenza and measles.
- Paroxysms may recur 50 or 60 times a day in pertussis, episodes begin after 2 weeks, may persist for 2 weeks with gradual improvement for 2 weeks; paroxysms at times in influenza.
- Severe in pertussis and influenza.
- Productive in pertussis; productive in about 1/3 in influenza; nonproductive in measles.
- Sputum copious and mucoid.
- Hemoptysis rare.
- Cough most common at night during early stages of pertussis.
- Persists during course of disease.

ASSOCIATED FINDINGS

- Fever, if any, slight in pertussis; an initial symptom in measles; a major symptom in influenza.
- Suffused conjunctivae, subcutaneous hemorrhages, periorbital edema and petechiae in pertussis; coryza, rhinorrhea and conjunctivitis in influenza and measles.
- Generalized aching in influenza.
- Cyanosis at times in pertussis; not observed in uncomplicated influenza or in measles.
- Diffuse rhonchi and occasional patches of moist rales at times in pertussis; absent in influenza unless complicated by pneumonia.
- Very marked leukocytosis with 70%-80% lymphocytes in pertussis; leukopenia frequent in influenza.
- Nasopharyngeal cultures will reveal *Bordetella pertussis* in pertussis.
- Koplik's spots appear within 2 days, rash in about 3 or 4 days after cough, in measles.

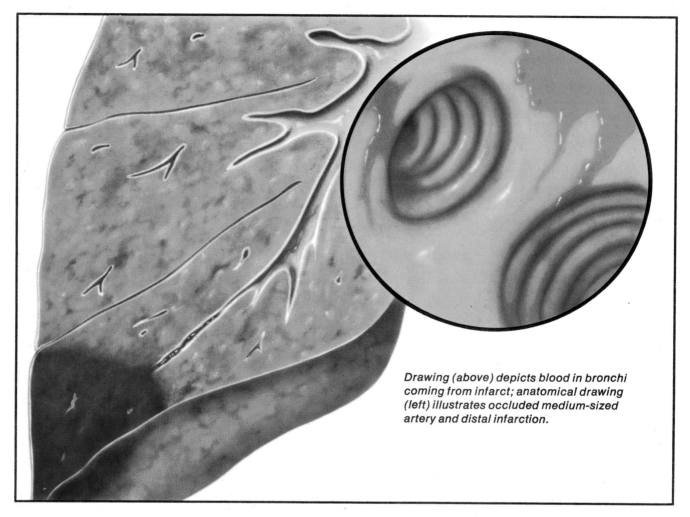

Drawing (above) depicts blood in bronchi coming from infarct; anatomical drawing (left) illustrates occluded medium-sized artery and distal infarction.

Cough due to
PULMONARY INFARCTION

Cough is not always present with pulmonary infarction; however, it can be an important initial symptom. If the postoperative patient or the patient with congestive heart failure or thrombophlebitis develops fever, tachycardia, pleuritic chest pain or hemoptysis, pulmonary infarction should be suspected. The picture is often incomplete and diagnosis can be missed.

- Onset of cough usually acute.
- Never paroxysmal; always mild to moderate.
- Usually nonproductive; when productive (in septic embolus), sputum is purulent.
- Hemoptysis common.
- Cough not affected by position or time of day.
- May last for several days to a week.

ASSOCIATED FINDINGS

- Fever most consistent symptom but not always present.
- Sharp and jabbing chest pain on breathing over infarct or supraclavicular fossa.
- Respiration rapid and shallow because of pain.
- Heart rate rapid.
- Pleural friction rub in about 25%; a patch of inspiratory moist rales more common.
- Signs of consolidation or pulmonary hypertension uncommon.
- Pleural effusion, if present, will not appear for days; may be serous or bloody; specific gravity and protein more like transudate than exudate.
- ECG may show shift of QRS axis to right, right bundle branch block or atrial fibrillation.
- Infarct not always visible on x-ray; may show wedge-shaped or patchy infiltrate after 12 to 24 hours; elevated diaphragm, with effusion at times.
- SGOT normal, except with acute right ventricular failure and hepatic congestion; serum bilirubin often rises after 3 or 4 days; serum LDH often begins to rise within 24 hours; peaks in 2 or 3 days; may last 10 days.
- Lung scan shows area of underperfusion.

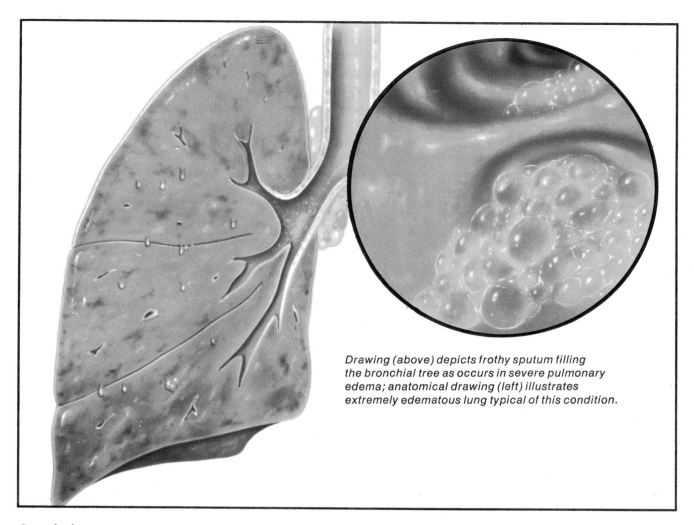

Drawing (above) depicts frothy sputum filling the bronchial tree as occurs in severe pulmonary edema; anatomical drawing (left) illustrates extremely edematous lung typical of this condition.

Cough due to
ACUTE PULMONARY EDEMA

Acute pulmonary edema, whether due to left ventricular failure or mitral stenosis, is usually characterized by intense dyspnea and cough. Cough can be the initial symptom when it awakens the patient from sleep. Occasionally, the cough can be dominant and obscure diagnosis.

- Onset acute or insidious; usually follows dyspnea.
- At first moderate; becomes persistent and even paroxysmal in severe cases.
- Usually nonproductive, productive in severe cases; when productive, sputum is copious and loose; frothy pink in very severe cases.
- Gross hemoptysis may occur with mitral stenosis.
- Cough most frequent on exertion or lying down.
- Lasts as long as edema — hours to days.

ASSOCIATED FINDINGS

- Fever absent.
- Severe dyspnea — exertional, orthopneic or paroxysmal nocturnal.
- Breathing rapid, shallow and labored.
- Perspiration marked.
- Skin cool and clammy; sometimes cyanotic.
- Tachycardia nearly always present; rhythm at times irregular due to atrial fibrillation.
- Profuse inspiratory fine and medium moist rales, more at bases than apices; often absent in chronic heart failure.
- Pulmonic component of second heart sound accentuated.
- Broad and sustained apical impulse, protodiastolic and often presystolic gallop, palpable pulsus alternans at times — with left ventricular failure.
- Apical diastolic rumble, often opening snap of mitral valve, sometimes xiphoid presystolic gallop — with mitral stenosis.
- Venous distention, hepatomegaly, pleural effusion, dependent edema at times — with right heart failure.
- X-ray shows prominent pulmonary vessels, usually cardiomegaly.

DIFFERENTIAL DIAGNOSIS OF COUGH

	Onset	Severity	Character	Productivity
CHRONIC BRONCHITIS	Insidious	Variable—mild in summer, severe in winter	Paroxysmal at times; more marked during night or on arising	Tenacious mucoid or mucopurulent sputum
THORACIC AORTIC ANEURYSM	Insidious	Never severe	Not paroxysmal; may be "brassy"	Usually nonproductive; when productive, mucopurulent sputum
BRONCHOGENIC CARCINOMA	Insidious	Mild at first, may become severe	Frequent but usually not paroxysmal	Mucoid or mucopurulent sputum; nonproductive at times
PULMONARY TUBERCULOSIS	Insidious	Becomes severe	Initially dry and hacking; frequent but usually not paroxysmal	Mucopurulent sputum (often green)
BRONCHIECTASIS	Insidious	Variable—mild, moderate or severe	Prolonged paroxysms; more marked on arising, lying or bending	Copious, purulent sputum; less tenacious than lung abscess
PNEUMOCOCCAL LOBAR PNEUMONIA	Usually insidious	Mild at first, becomes severe	Paroxysmal at times	Sputum tenacious, mucoid and "rusty"; becomes loose and mucopurulent

Presence of Blood	Duration	Associated Signs and Symptoms	Key Laboratory Data
Hemoptysis uncommon	Chronic and persistent (at least 3 months)	Progressive exertional dyspnea; malaise slight; fever absent; roughened breath sounds; transient expiratory wheezes; fine moist rales; paranasal sinusitis and posterior rhinitis often present	X-ray normal or reveals increased hilar markings, emphysematous changes
Hemoptysis rare	Chronic and persistent	Chest pain; dysphagia; inspiratory stridor; basal diastolic murmur; tracheal tug; superior vena caval syndrome; sometimes signs of Marfan's syndrome	Fusiform or saccular dilatation of aorta on x-ray; positive serologic test for syphilis
Blood-streaked sputum; frank hemoptysis common	Chronic and persistent	Weakness; malaise; chest pain; weight loss; anemia; hoarseness; localized wheezing; pulmonary osteoarthropathy; enlarged nodes	X-ray—new density or mediastinal widening, without calcification; 50% can be diagnosed by bronchoscopy
Hemoptysis common; massive hemorrhage occasional	Chronic and persistent	Fever; night sweats; weakness; weight loss; fine and medium moist rales, mostly over upper lung; cavernous breath sounds over cavities	Tuberculin test positive; sputum or gastric culture positive; x-ray will show patchy infiltrates with areas of fibrosis or cavitation
Hemoptysis common; massive hemorrhage rare	Chronic and persistent	Fever absent; medium or coarse moist rales and rhonchi; amphoric breath sounds over dilated bronchi; clubbing of fingers may occur; intermittent pneumonias	Sputum layers out on standing; pus cells are polys; saccular dilatations on bronchogram diagnostic
Rusty sputum (degraded blood); frank hemoptysis uncommon	8 to 10 days if untreated	Marked fever and prostration; headache; dyspnea; cyanotic at times; pleuritic chest pain; signs of consolidation and moist rales over involved lobe; pleural friction rub	X-ray shows opacification of involved lobe; pneumococci often cultured from sputum and blood

(Continued on next page.)

DIFFERENTIAL DIAGNOSIS OF COUGH (continued)

	Onset	Severity	Character	Productivity
KLEBSIELLA PNEUMONIA	Usually acute	Severe from onset	Frequently paroxysmal	Copious, tenacious, mucoid sputum
MYCOPLASMA PNEUMONIA	Acute	Often severe	Paroxysmal much of the time	Becomes productive; mucoid or mucopurulent sputum
LUNG ABSCESS	May be acute or insidious	Becomes severe	Frequent but not usually paroxysmal	Copious, tenacious, purulent sputum; often fetid
ACUTE TRACHEOBRONCHITIS	Acute in influenza and measles, insidious in pertussis	Severe in pertussis and influenza, variable in measles	Paroxysmal in pertussis; paroxysmal at times in influenza; not usually in measles	Copious, mucoid sputum in pertussis; productive at times in influenza; not usually in measles
PULMONARY INFARCTION	Usually acute	Mild to moderate	Never paroxysmal	Usually non-productive; purulent sputum with septic embolus
PULMONARY EDEMA	Acute or insidious	Moderate to severe	Paroxysmal at times; most frequent on exertion or lying down	Usually non-productive; frothy pink sputum in severe cases

Presence of Blood	Duration	Associated Signs and Symptoms	Key Laboratory Data
Brick-red or brown and ropy sputum; hemoptysis common, at times marked	Course of disease (usually several weeks)	Fever, chill and prostration; headache, delirium; dyspnea and cyanosis; pleuritic chest pain; signs of consolidation and moist rales over several lobes; disease of alcoholics, malnourished or debilitated	X-ray will show opacification of one or more lobes, cavitation; gram-negative rods in stain of sputum
Blood-streaked sputum at times; no frank hemoptysis	Course of disease (usually several weeks)	Chilly sensations, marked fever, no prostration; headache; substernal chest pain; patchy fine and medium rales; sore inflamed throat with scanty exudate	X-ray will show scattered densities, more than expected by physical signs; cold agglutinins in about 50%
Bloody sputum and frank hemoptysis common; massive hemorrhage rare	Persists for weeks if untreated	Fever, malaise and pleuritic chest pain; inspiratory rales and dullness to percussion over lesion; clubbing of fingers may develop; gingivodental disease frequent	X-ray shows round or segmental infiltrate, often with air fluid level
Hemoptysis rare	Persists for a week or two in influenza and measles; for 6 weeks in pertussis	Fever (slight in pertussis); generalized aching in influenza; suffused conjunctivae and periorbital edema in pertussis; coryza, rhinorrhea and conjunctivitis in influenza and measles; Koplik's spots precede rash in measles	Very marked leukocytosis in pertussis; leukopenia often in influenza; positive culture in pertussis
Hemoptysis common	Several days to a week	Fever; pleuritic chest pain; dyspnea; tachycardia; patch of inspiratory moist rales; pleural friction rub; signs of consolidation	X-ray shows wedge-shaped or patchy infiltrate in 12 to 24 hours in most cases; SGOT usually normal; LDH and bilirubin elevated; ECG may show shift of axis to right
Frank hemoptysis with mitral stenosis	Lasts as long as edema; hours to days	Dyspnea and perspiration; fever absent; profuse inspiratory fine and medium moist rales; tachycardia; gallop rhythm or mitral diastolic rumble	X-ray shows prominent pulmonary vessels, usually cardiomegaly

PULMONARY RALES

Although auscultation of pulmonary rales has been de-emphasized somewhat in recent years in favor of chest x-ray findings, pulmonary rales can serve as an important physical sign in differential diagnosis.

Rales are caused by a variety of disorders of the lungs and circulation. When present, a careful analysis of their character and location, in conjunction with other signs and symptoms, will facilitate diagnosis.

The nomenclature of pulmonary rales has not been standardized. In this presentation, we define rales as either moist or dry and further describe their characteristics in eleven conditions in which they are an important diagnostic sign.

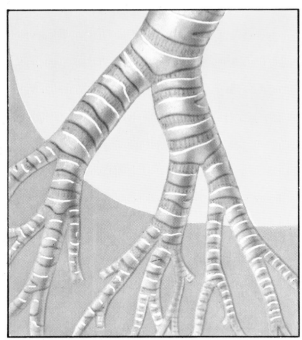

SITES OF ORIGIN OF <u>DRY RALES</u>
☐ Sonorous (rhonchi)
☐ Sibilant (wheezes)

SITES OF ORIGIN OF <u>MOIST RALES</u>
☐ Coarse ☐ Medium ■ Fine

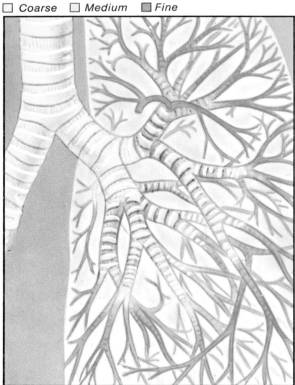

DRY RALES

Adventitious sounds *of long duration* produced by movement of air through partially obstructed bronchi:

SONOROUS (rhonchi)

- Groaning sounds in the large bronchi.
- Expiratory, may be inspiratory at times.
- Occasionally reduced by coughing.
- Vibrations of chest wall easily felt on palpation.

SIBILANT (wheezes)

- High-pitched whistling sounds in smaller bronchi.
- Expiratory in most, rarely inspiratory.
- Usually not reduced by coughing.
- Can sometimes be heard at a distance.

MOIST RALES

Adventitious sounds *of short duration* produced by the movement of air: 1) through collections of fluids, or 2) between adherent surfaces of the bronchial tree:

FINE (crepitant)

- Crackling sounds in distal portion of bronchial tree.
- Typically in showers at end of inspiration.
- Usually not reduced by coughing.

MEDIUM (subcrepitant)

- Bubbling sounds between distal portion of bronchial tree and large bronchi.
- Mostly late inspiratory.
- Usually not reduced by coughing.

COARSE (tracheal)

- Gurgling sounds in trachea and larger bronchi.
- Both inspiratory and expiratory. On early inspiration in terminal cases (so-called "death rattle").
- Sometimes reduced by coughing.

Some moist rales may have additional characteristics requiring further descriptive nomenclature:

- *Consonating*—loud and snapping due to resonance through consolidated area.
- *Atelectatic*—fine or medium rales at bases of the lung, which disappear after a few deep breaths or coughs.
- *Post-tussive*—a shower of fine rales heard only after coughing; during early inspiration.

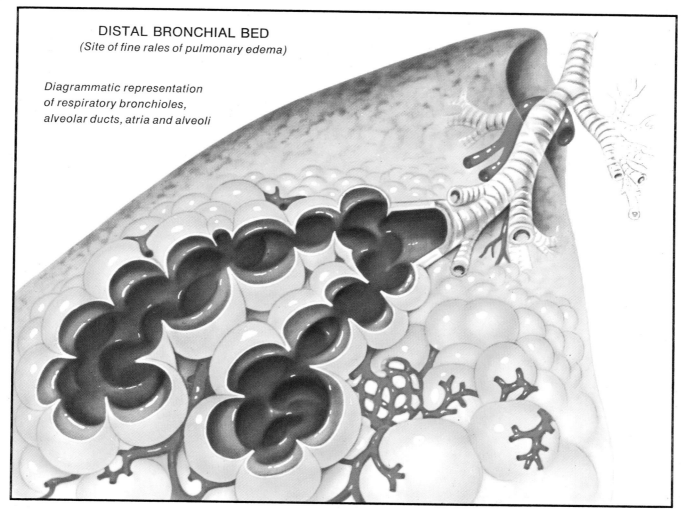

DISTAL BRONCHIAL BED
(Site of fine rales of pulmonary edema)

Diagrammatic representation of respiratory bronchioles, alveolar ducts, atria and alveoli

Pulmonary rales due to
ACUTE PULMONARY EDEMA

Rales are usually associated with acute episodes of dyspnea in pulmonary edema. The transudation of fluid from engorged pulmonary capillaries into pulmonary alveoli and small bronchi, whether due to left ventricular failure or mitral stenosis, produces moist rales. Dry sibilant rales may also be present.

- Fine or medium moist (crepitant) in milder cases; consonating with consolidation of lower lobes; coarse and bubbling rales in severe cases.
- Maximum at end of inspiration; inspiratory and expiratory in severe cases.
- Tend to be restricted to bases of both lungs; numerous and extensive in severe cases.
- More numerous on side on which patient lies.
- Usually occur with, and persist as long as, acute dyspnea; may be chronic.
- Few diffuse sibilant rales (wheezes) often present; when they obscure moist rales in cardiac asthma, condition can be misdiagnosed as bronchial asthma.

ASSOCIATED FINDINGS

- Perspiration marked; no fever; skin cool and clammy; sometimes cyanotic.
- Breathing rapid, shallow and labored; severe dyspnea — exertional, orthopneic or paroxysmal nocturnal.
- Frequent nonproductive cough, at times paroxysmal; productive with copious loose sputum in severe cases; frothy pink preterminally.
- Tachycardia nearly always present; rhythm irregular at times due to atrial fibrillation.
- Pulmonic component of second heart sound loud.
- Broad and sustained apical impulse, protodiastolic and often presystolic gallop, palpable pulsus alternans at times — with left ventricular failure.
- Apical diastolic rumble, often opening snap of mitral valve, xiphoid presystolic gallop at times — with mitral stenosis.
- Venous distention, hepatomegaly, pleural effusion, dependent edema — with right heart failure.
- X-ray shows prominent pulmonary vessels, usually cardiomegaly.

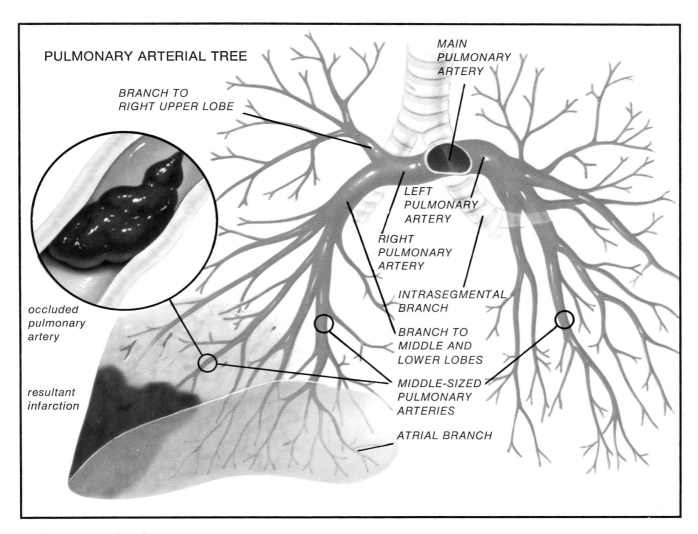

Pulmonary rales due to PULMONARY INFARCTION

Although rales are present in only about 25% of cases of pulmonary infarction, they can be of diagnostic value since any or all of the typical findings in this condition may be absent and diagnosis overlooked. Therefore, pulmonary infarction must be considered when moist rales are present: postoperatively, in congestive heart failure, with any pulmonary vascular dysfunction or in the presence of thrombophlebitis.

- Fine or medium moist (crepitant); mainly at end of inspiration.
- Localized to small area, but may be extensively distributed with multiple infarcts.
- Unlike atelectatic rales common in postoperative and immobilized patients, pulmonary infarct rales do not clear after deep breathing or coughing.
- Usually occur about a day after onset of condition; may persist for several days to a week.

ASSOCIATED FINDINGS

- May be preceded by symptoms of embolism (midline chest pain, dyspnea and/or tachycardia).
- Respiration rapid and shallow because of pain.
- Cough usually nonproductive; when productive (in septic embolus), sputum is purulent; hemoptysis common.
- Sharp and jabbing pain on breathing, over infarct or supraclavicular fossa, an early symptom.
- Fever most consistent symptom; not always present.
- Pleural friction rub often; signs of consolidation occasional, in same area as rales.
- Pleural effusion, if present, will not appear for days; may be serous or bloody — protein and specific gravity more like transudate than exudate.
- ECG may show shift of QRS axis to right, right bundle branch block or atrial fibrillation.
- Infarct not always visible on x-ray — may show wedge-shaped or patchy infiltrate after 12 to 24 hours, elevated diaphragm or evidence of effusion; lung scan shows area of underperfusion.
- SGOT normal, except with acute right ventricular failure and hepatic congestion; serum bilirubin often rises after 3 or 4 days; serum LDH may begin to rise within 24 hours, peaks in 2 or 3 days, may last 10 days.

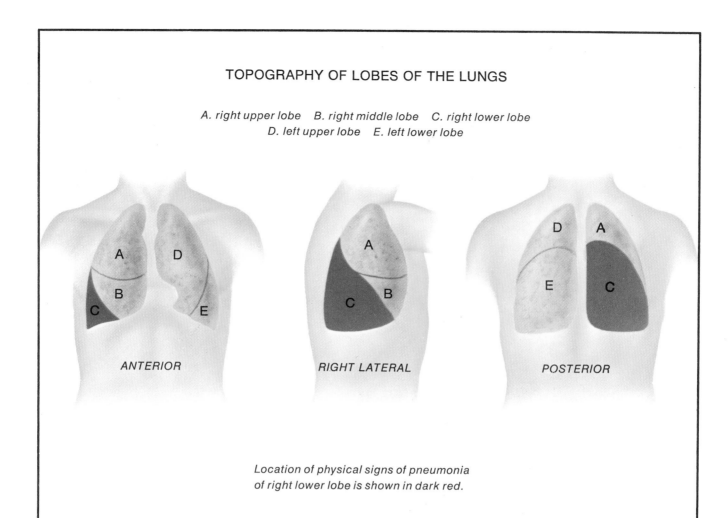

TOPOGRAPHY OF LOBES OF THE LUNGS

A. right upper lobe B. right middle lobe C. right lower lobe
D. left upper lobe E. left lower lobe

ANTERIOR RIGHT LATERAL POSTERIOR

Location of physical signs of pneumonia of right lower lobe is shown in dark red.

Pulmonary rales due to LOBAR PNEUMONIA

In lobar pneumonia, rales usually occur with inflammatory consolidation of a lobe due to exudate in the alveoli and small bronchi and persist as long as the lobe is aerated. With early diagnosis and use of antibiotics, typical consolidation may not occur; but rales will nonetheless be present. Pneumococcal lobar pneumonia is usually preceded by an upper respiratory infection.

- Fine or medium moist (crepitant) early; coarse, often consonating, during resolution; mainly at end of inspiration.
- Initially may be restricted to small area high in axilla and easily missed; distribution extends over entire involved lobe with consolidation.
- Numerous during early stages and resolution; less numerous with consolidation.
- Usually occur on first day of illness; can persist as long as a week after temperature falls.

ASSOCIATED FINDINGS

- Marked fever and prostration; often follows severe chill.
- Headache, even delirium, in severely ill; skin usually warm and flushed.
- Severe pleuritic pain over involved lobe with tenderness and hyperesthesia of chest wall.
- Cough initially slight and nonproductive; becomes severe and productive of purulent sputum.
- Severe dyspnea with rapid, shallow breathing and expiratory grunting; cyanosis at times.
- Slight change in whispered voice may be only auscultatory finding in early stages; later, pleural friction rub and signs of consolidation.
- Signs of fluid common, empyema in 5%.
- Herpetic blisters around mouth in some.
- Abdominal distention frequent.
- WBC elevated; increase in polymorphonuclear leukocytes.
- Sputum smear may show encapsulated gram-positive cocci; pneumococcus often cultured from both sputum and blood.
- X-ray shows opacification of consolidated lobe.

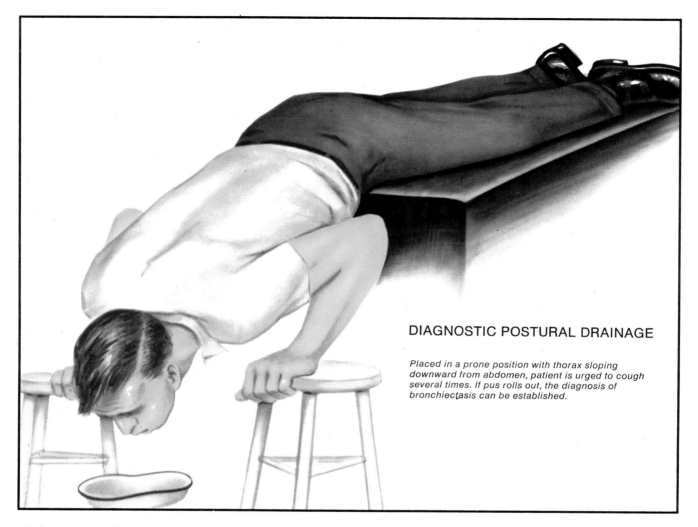

DIAGNOSTIC POSTURAL DRAINAGE

Placed in a prone position with thorax sloping downward from abdomen, patient is urged to cough several times. If pus rolls out, the diagnosis of bronchiectasis can be established.

Pulmonary rales due to BRONCHIECTASIS

Rales are usually present in bronchiectasis because of exudate in the bronchi. They are unaffected by the characteristic cough of the condition. Bronchiectasis should be suspected when persistent rales are present at the lung bases in a resonant chest, or when a chronic productive cough is associated with pneumonia and hemoptysis. The condition usually follows necrotizing pneumonia or infection of congenital cysts.

- Fine or medium moist (crepitant); coarse and consonating when dilatation is marked.
- Mainly inspiratory.
- In one or several areas of lung, mostly at bases.
- Positioning of patient to improve drainage may cause partial clearing of moist rales.
- Usually chronic.
- Sonorous rales (rhonchi) may be heard in same area as moist rales.
- Diffuse sibilant rales (wheezes) may be present with associated bronchitis.

ASSOCIATED FINDINGS

- Chronic productive cough with copious, tenacious and purulent sputum; nonproductive cough in "dry" bronchiectasis.
- Hemoptysis often present; massive hemorrhage rare.
- Dyspnea and cyanosis absent, except with chronic bronchitis and emphysema.
- Intermittent episodes of pneumonia with consolidation, fever and pleuritic pain.
- Occasionally amphoric breath sounds over large cystic dilatations.
- Clubbing of fingers occasionally seen.
- Paranasal sinusitis frequent.
- Anemia sometimes present.
- Sputum "layers" on standing (mucus on top, pus below); pus cells are polys, not eosinophils; culture may show various bacteria.
- X-ray shows increased peribronchial markings in basilar segments of lower lobes.
- Saccular dilatations on bronchogram are diagnostic.

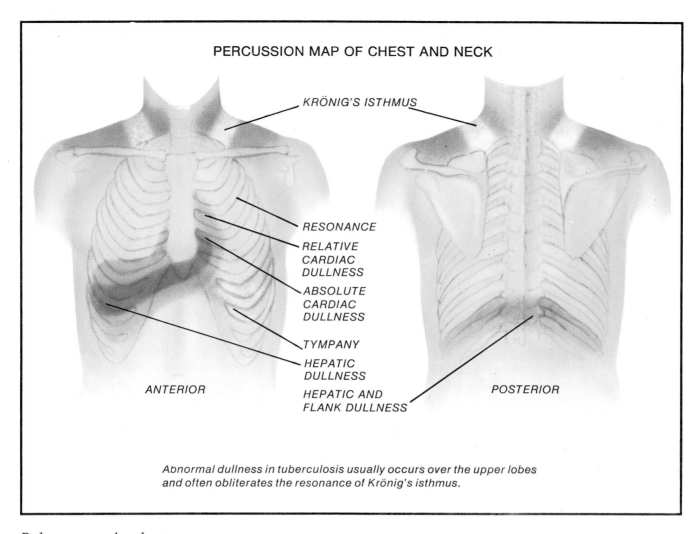

Abnormal dullness in tuberculosis usually occurs over the upper lobes and often obliterates the resonance of Krönig's isthmus.

Pulmonary rales due to PULMONARY TUBERCULOSIS

Rales and a chronic productive cough are prominent findings in moderate and advanced adult-type pulmonary tuberculosis, but are usually not present with mild infections. Rales and chronic cough are *not* features of primary tuberculosis in children or of miliary tuberculosis.

- Fine or medium moist; post-tussive with moderate lesion; persistent with larger lesion; coarse and medium moist with far-advanced lesion.
- Consonating and sharp with consolidation; consonating and metallic near cavitation.
- At onset of inspiration when post-tussive; mostly end-inspiratory with larger lesion.
- Usually in small patches at apices when post-tussive; more extensive with larger lesion, but still usually restricted to upper chest.
- Usually persist as long as lesion remains active.
- Sometimes also sibilant rales (wheezes) with larger lesion.

ASSOCIATED FINDINGS

- Suppression of menses frequent.
- Cough nearly always present; initially dry and hacking, later productive of mucopurulent sputum.
- Commonly fever (can be high without acute toxicity), night sweats, malaise, weakness, easy fatigability, weight loss.
- Dullness to percussion with extensive involvement; narrowing of band of supraclavicular percussive resonance with apical disease; cavernous breath sounds and tympanitic percussion note over cavity when cavity walls are rigid, near periphery and communicate with bronchus.
- With percussion over cavity, "cracked-pot" resonance may be heard with stethoscope over open mouth.
- Smear and culture of sputum, or culture of gastric contents diagnostic.
- X-ray shows patchy infiltrates, at times with areas of fibrosis or cavitation and compensatory emphysema.
- Tuberculin test almost always positive in adult-type pulmonary tuberculosis.

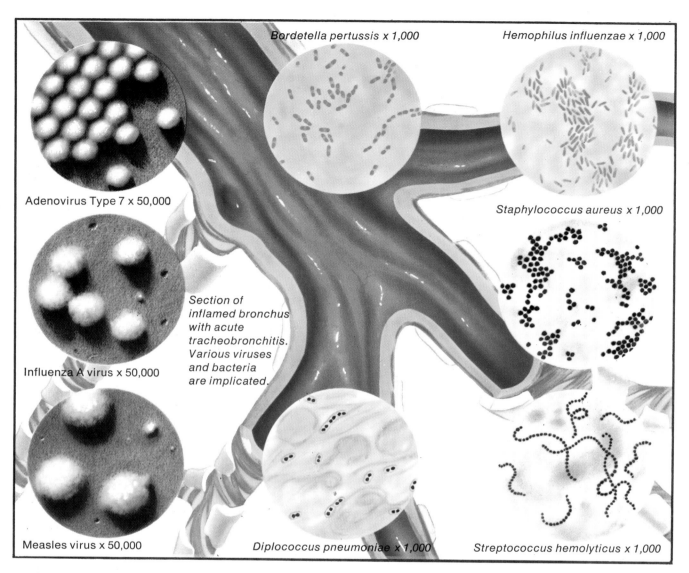

Pulmonary rales due to
ACUTE TRACHEOBRONCHITIS

Dry rales, nonproductive cough and fever are characteristic in the early stages of acute tracheobronchitis; moist rales, productive cough and fever occur in the later stages. The acute inflammation of the bronchi is produced by viruses, bacteria or external irritants. It is more likely to complicate viral upper respiratory infections in children than in adults, and is frequently a feature of measles, influenza and pertussis.

- Sonorous and sibilant dry (rhonchi and wheezes) in early stages; fine or medium moist (crepitant) in later stages; coarse as condition progresses.
- Dry rales predominantly expiratory; moist rales inspiratory.
- Diffuse and bilateral; localized moist rales may be due to complicating pneumonia.
- May disappear on coughing or be persistent.

ASSOCIATED FINDINGS

- Cough productive in pertussis; productive in about 1/3 in influenza; nonproductive in measles and most other viral upper respiratory infections.
- Hemoptysis uncommon.
- Fever a major symptom in viral tracheobronchitis, but minimal in pertussis.
- Suffused conjuctivae, subcutaneous hemorrhages, periorbital edema and petechiae in pertussis; coryza, rhinorrhea, conjunctivitis in influenza and measles.
- Cyanosis at times in pertussis; not in measles or uncomplicated influenza.
- "Whoop" may be followed by vomiting in pertussis.
- Very marked leukocytosis with 70% to 80% lymphocytes in pertussis; leukopenia often in influenza.
- Nasopharyngeal cultures reveal *Bordetella pertussis* in pertussis.
- Koplik's spots appear within 2 days, rash about 3 or 4 days after cough, in measles.
- Chest x-ray normal.
- Bronchoscopy shows reddened, inflamed trachea and bronchi.

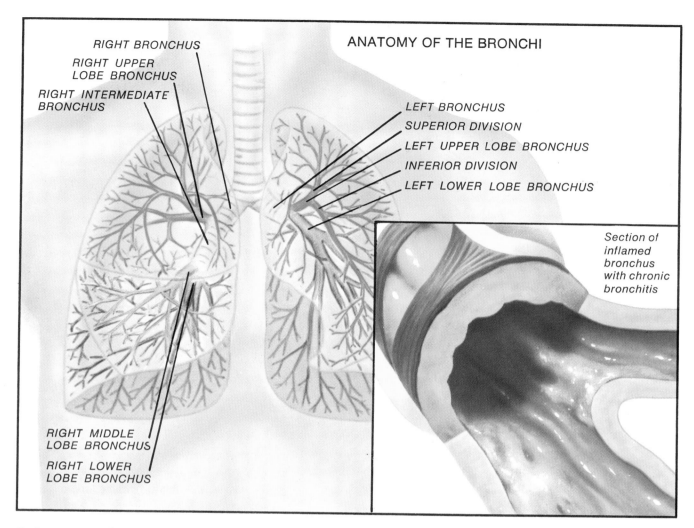

Pulmonary rales due to
CHRONIC BRONCHITIS

Dry rales and chronic productive cough are characteristic in chronic bronchitis. This condition is probably the most common cause of sonorous and sibilant rales. It is commonly associated with pulmonary emphysema, usually occurring in people over 40 and four times more frequently in men than in women. It may also be prominent in patients with cystic fibrosis of the pancreas. Superimposed acute bronchitis may intensify all symptoms.

- Sonorous and sibilant dry (rhonchi and wheezes).
- Predominantly expiratory; inspiratory at times.
- Diffuse and bilateral.
- In severe cases, coarse rhonchi can be felt over both lower lung fields.
- Pattern and number can be changed by coughing.
- Persistent and chronic; more marked in winter.
- Few inspiratory fine or medium moist rales, usually at bases; if numerous or chronic, bronchiectasis should be suspected.

ASSOCIATED FINDINGS

- Productive cough with tenacious, mucoid or mucopurulent sputum; hemoptysis uncommon.
- Progressive exertional dyspnea.
- Fever absent; malaise slight except with superimposed acute bronchitis.
- Breath sounds roughened.
- No signs of consolidation.
- Paranasal sinusitis and posterior rhinitis often associated, especially in cigarette smokers.
- Many organisms can be cultured; pneumococcus and *Hemophilus influenzae* most common.
- Chest x-ray may be normal or show increased hilar markings; may reveal hyperlucency and depressed diaphragm in presence of emphysema.
- Bronchogram shows cylindrical bronchial dilatation, not saccular as in bronchiectasis.
- Bronchoscopy reveals normal or slightly reddened bronchial mucosa.

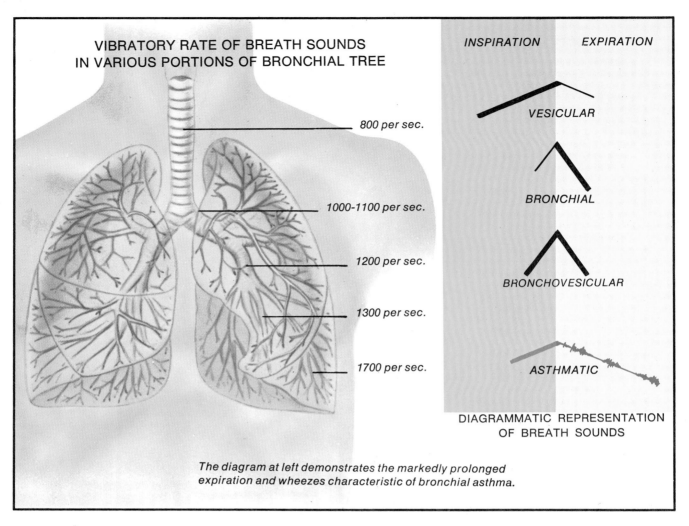

The diagram at left demonstrates the markedly prolonged expiration and wheezes characteristic of bronchial asthma.

Pulmonary rales due to BRONCHIAL ASTHMA

Attacks of bronchial asthma are characterized by dry rales with dyspnea; showers of fine moist rales, however, can also appear at the end of an attack. Cardiac asthma can easily be misdiagnosed as bronchial asthma; the presence of apical gallop rhythm usually establishes the differential diagnosis.

- Sonorous and sibilant dry (rhonchi and wheezes).
- Predominantly expiratory; inspiratory at times.
- The higher pitched the wheezing, the more severe the attack and obstruction.
- Extensively distributed throughout the chest.
- At times may be heard at a distance.
- May be chronic or present only during attacks.
- Showers of fine moist rales (crepitant) at end of attack, mostly at bases; often disappear after productive cough.

ASSOCIATED FINDINGS

- Attacks often preceded by itching of chest or sneezing.
- Paroxysmal wheezing and dyspnea often accompanied by tightness and burning in chest.
- Jerky, nonproductive cough usually follows dyspnea; improvement occurs when cough becomes productive.
- Perspiration profuse; no fever.
- In brief attacks, patient sits up, may lean over table or chair back; in prolonged attacks, patient lies prostrate, uses accessory muscles less, becomes cyanotic.
- Breathing labored during attacks with short inspiration and greatly prolonged expiration; breathing becomes rapid despite prolonged expiration in severe attacks; expiratory breath sounds longer and higher pitched than in bronchial breathing.
- Chest hyperresonant; lung bases low; vocal and tactile fremitus diminished.
- Sometimes inspiratory retraction of lower ribs from contraction of flat diaphragm.
- Sputum mucoid and contains many eosinophils.
- X-ray normal or shows hyperlucent lung fields.

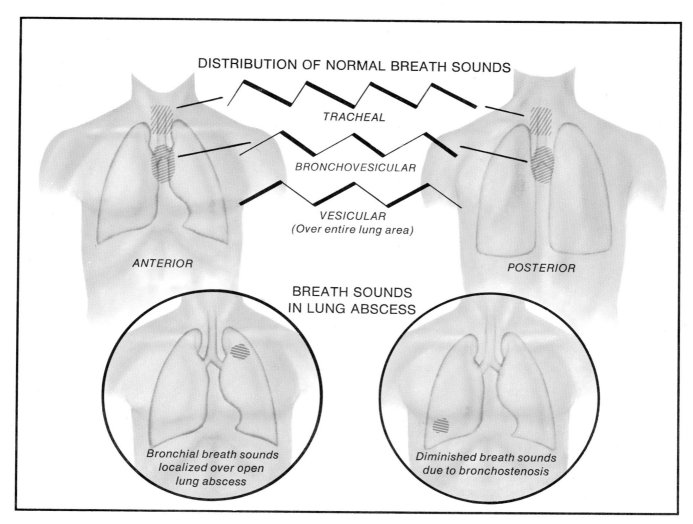

Pulmonary rales due to
LUNG ABSCESS

Moist rales are heard early in the course of lung abscess. Although the findings of rales, cough, malaise and fever of this condition may suggest acute bronchitis or pneumonia, the pleuritic chest pain of lung abscess separates it from bronchitis. The character of sputum, x-ray appearance and persistence of signs and symptoms separate it from pneumonia. Lung abscess, a suppurative pneumonic process with resultant necrosis, may follow aspiration of septic material or an obstruction to bronchial drainage.

- Medium moist (subcrepitant); consonating with significant consolidation.
- Predominantly inspiratory.
- Localized to area of abscess.
- Occur early, before cavitation appears; may last for few weeks or even months.
- Localized sibilant dry rales (wheezes) with bronchostenosis.

ASSOCIATED FINDINGS

- Fever and malaise always present.
- Nonproductive cough initially; becomes productive with copious sputum, often fetid and/or bloody.
- Sharp and jabbing pleuritic chest pain on breathing for several days during acute stage.
- Breathing shallow, proportional to pain.
- Pleural friction rub often present.
- Amphoric breath or voice sounds not usually heard because cavity is thin walled.
- Clubbing of fingers may occur with subacute or chronic lung abscess; usually develops late, 1 to 2 weeks at earliest.
- Gingivodental disease with foul breath frequent.
- Obstruction of bronchial drainage often present from stricture, carcinoma or inspissated mucus.
- Various organisms (including spirochetal and fusiform) may be cultured.
- X-ray will show round or segmental infiltrate; often, but not always, with air fluid level.
- WBC often elevated to 20,000 to 30,000.

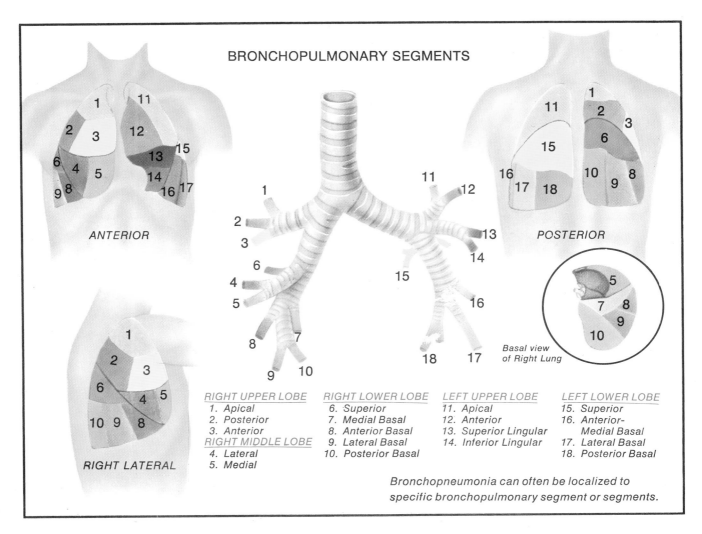

BRONCHOPULMONARY SEGMENTS

RIGHT UPPER LOBE
1. Apical
2. Posterior
3. Anterior

RIGHT MIDDLE LOBE
4. Lateral
5. Medial

RIGHT LOWER LOBE
6. Superior
7. Medial Basal
8. Anterior Basal
9. Lateral Basal
10. Posterior Basal

LEFT UPPER LOBE
11. Apical
12. Anterior
13. Superior Lingular
14. Inferior Lingular

LEFT LOWER LOBE
15. Superior
16. Anterior-Medial Basal
17. Lateral Basal
18. Posterior Basal

Bronchopneumonia can often be localized to specific bronchopulmonary segment or segments.

Pulmonary rales due to BRONCHOPNEUMONIA

Moist rales are a typical finding in bronchopneumonia and may be the only clue to diagnosis of this condition. Dry rales can also be present if there is concomitant pulmonary disease. Bronchopneumonia usually complicates preexisting pulmonary or debilitating systemic disease, but streptococci or pneumococci can produce severe *primary* bronchopneumonia.

- Fine or medium moist (crepitant).
- Predominantly inspiratory.
- Scanty at bases at several distinct areas.
- Can be diffuse and scattered.
- May change markedly from day to day.
- Occur early and persist during course of disease.
- Sonorous and sibilant rales (rhonchi and wheezes) with concomitant chronic bronchitis and emphysema.

ASSOCIATED FINDINGS

- Prostration often marked.
- Fever usual, but may be absent in severely debilitated; differs from fever of lobar pneumonia — has wide diurnal swings.
- Dyspnea, when present, usually due to underlying disease.
- Cough productive of mucopurulent sputum.
- Cough can be absent in severely debilitated.
- Pleuritic pain and friction rub at times — simulating pulmonary infarct.
- Signs of consolidation present at times; borders are indistinct.
- Empyema often develops within 48 hours in primary streptococcal or pneumococcal bronchopneumonia.
- Organisms identified by sputum and often blood culture.
- X-ray shows one or more patchy infiltrates with ill-defined borders; not as dense as lobar consolidation.

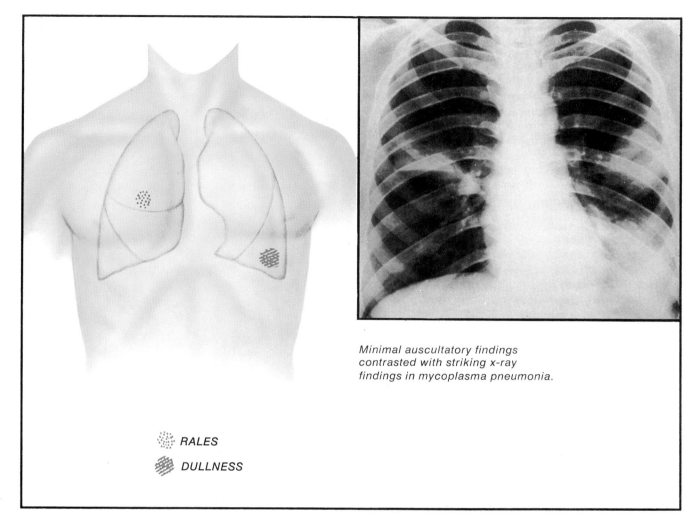

Minimal auscultatory findings contrasted with striking x-ray findings in mycoplasma pneumonia.

RALES

DULLNESS

Pulmonary rales due to
MYCOPLASMA PNEUMONIA

Moist rales may be the only auscultatory finding in mycoplasma pneumonia; sometimes, however, dry rales may also be present. Mycoplasma pneumonia occurs sporadically. It is usually preceded by an upper respiratory infection, and is most common in school-age children. It is seldom seen in people over the age of 50. Chest x-ray is the most important clue to diagnosis.

- Fine or medium moist (crepitant).
- Coarse during resolution.
- Predominantly inspiratory.
- Localized to involved area.
- Occur early and persist during course of disease.
- Sonorous dry rales (rhonchi) may also be present in same area; intensified by forced expiration.

ASSOCIATED FINDINGS

- Headache and malaise may be severe, but no delirium or prostration.
- Marked fever and chilly sensations.
- Nonproductive cough at first; becomes productive with mucoid or mucopurulent sputum, often blood-streaked; paroxysmal much of the time.
- Substernal rather than pleuritic pain; aggravated by coughing.
- Dyspnea and cyanosis rare.
- Throat usually sore, nasopharyngeal membrane inflamed with scanty exudate.
- Pleural friction rub or effusion rare.
- Relatively slow heart rate.
- Little if any elevation of WBC; ESR almost always elevated.
- X-ray shows scattered mottled densities mostly in region of hila and lower lobes, may migrate; more marked than expected from physical exam.
- Cold agglutinins can be demonstrated in about 50% of patients; special serologic procedures will differentiate from pneumonia of Q fever or psittacosis.

DIFFERENTIAL DIAGNOSIS OF PULMONARY RALES

	Type	Location	Duration
ACUTE PULMONARY EDEMA	Fine or medium moist; coarse and bubbling in severe; often scattered wheezes	Lung bases; diffuse when severe	As long as acute dyspnea; may be chronic
PULMONARY INFARCTION	Fine or medium moist	Area of infarct	Begin after first day; last about a week
LOBAR PNEUMONIA	Fine or medium moist early; coarse during resolution	Over involved lobe	Begin on first day; can last a week after fever
BRONCHIECTASIS	Fine or medium moist; can be coarse and consonating; reduced by postural drainage; rhonchi also present	One or several areas, usually at bases	Usually chronic
PULMONARY TUBERCULOSIS	Fine or medium moist; some post-tussive, some persistent; coarse with larger lesion; wheezes at times	Small areas at apices, or upper lungs	Persist as long as lesion remains active
ACUTE TRACHEOBRONCHITIS	Rhonchi and wheezes early; fine or medium moist later; coarse as condition progresses; may clear on coughing	Diffuse and bilateral	Persist as long as acute infection

Associated Signs and Symptoms	Key Laboratory Data
Dyspnea; orthopnea; nonproductive cough; marked perspiration; rapid, shallow breathing; tachycardia; loud P_2; gallop rhythm; opening snap (in mitral stenosis); venous distention; hepatomegaly	X-ray shows enlarged pulmonary vessels, usually cardiomegaly
Nonproductive cough; pleuritic chest pain; fever; hemoptysis; rapid, shallow breathing; pleural friction rub; signs of consolidation sometimes present; pleural effusion	Shift of QRS axis to the right at times; infiltrate on x-ray; elevated LDH and bilirubin
Fever; chill; prostration; headache; pleuritic chest pain; productive cough; dyspnea; rapid, shallow breathing; cyanosis at times; pleural friction rub; signs of consolidation; signs of fluid; herpetic blisters around mouth	WBC elevated; causative organism on smear of sputum; x-ray shows consolidated lobe
Chronic productive cough; purulent sputum; hemoptysis at times; amphoric breath sounds over large cystic dilatations at times; clubbing of fingers sometimes present; intermittent superimposed pneumonias; sputum "layers" on standing	Saccular dilatations in bronchogram
Chronic productive cough; fever; night sweats; malaise; dullness to percussion with larger lesion; sometimes limited to supraclavicular area; cavernous breath sounds over cavity	X-ray shows patchy infiltrates with fibrosis or cavitation; smear or culture of sputum diagnostic
Fever (minimal in pertussis); productive cough (except in measles); hemoptysis uncommon; coryza, rhinorrhea, conjunctivitis in measles, influenza and other viral conditions; suffused conjunctivae in pertussis, Koplik's spots and rash in measles	Leukocytosis in pertussis; leukopenia in influenza and other viral conditions; reddened bronchi on bronchoscopy

(Continued on next page.)

DIFFERENTIAL DIAGNOSIS OF PULMONARY RALES (continued)

	Type	Location	Duration
CHRONIC BRONCHITIS	Rhonchi and wheezes; few fine or medium moist at bases; pattern and number changed by coughing	Diffuse and bilateral	Persistent and chronic; more marked in winter
BRONCHIAL ASTHMA	Rhonchi and wheezes; showers of fine moist at end of attack	Diffuse and bilateral	Usually during attacks; may be chronic
LUNG ABSCESS	Medium moist; at times consonating; localized wheeze with bronchostenosis	Area of abscess	Occur early; may last weeks or months
BRONCHOPNEUMONIA	Fine or medium moist	Scanty at bases or in several areas; may change from day to day	Occur early; persist during disease
MYCOPLASMA PNEUMONIA	Fine or medium moist; coarse during resolution; rhonchi may also be present	Involved area or areas	Occur early; persist during disease

Associated Signs and Symptoms	Key Laboratory Data
Chronic productive cough; progressive dyspnea; no fever; hemoptysis uncommon; breath sounds roughened; no signs of consolidation; sinusitis and rhinitis often present	X-ray may be normal or increased hilar markings; bronchogram shows cylindrical bronchial dilatation
Itching and tightness of chest; wheezing; jerky nonproductive cough; profuse perspiration; cyanosis; labored breathing; prolonged expiration higher pitched than bronchial breathing; chest hyperresonant; lung bases low	Sputum mucoid with many eosinophils; x-ray may show hyperlucent lung fields
Fever; malaise; nonproductive cough becomes productive; copious sputum often fetid or bloody; pleuritic chest pain; breathing shallow; friction rub; clubbing of fingers at times; gingivitis	X-ray shows round or segmental infiltrate, often with fluid level
Fever with wide diurnal swings; prostration marked; productive cough; pleuritic pain at times; signs of consolidation at times, with indistinct borders; friction rub at times	X-ray shows one or more patchy infiltrates with ill-defined borders; organisms often cultured from sputum and blood
Fever; chilly sensations; headache; nonproductive cough becomes productive; mucoid sputum, often bloody; substernal pain; sore throat; cyanosis rare; pleural friction rub of effusion rare; slow heart rate	X-ray shows scattered mottled densities — may migrate; cold agglutinins in about 50%

PLEURAL EFFUSION

Accumulation of fluid in the pleural space is termed effusion, whether the fluid occurs in response to inflammation — exudation; or to elevated vascular pressure — transudation.

Accumulation of whole blood (hemothorax) or chyle (chylothorax) in the pleural space is not effusion. These conditions are usually produced by trauma.

A careful analysis of the character, location and specific features of pleural effusion in conjunction with other signs and symptoms will facilitate diagnosis.

In this presentation, general characteristics of pleural effusion are discussed and specific features of pleural fluid collections in eleven important disease entities are described and compared in tabular form.

General Characteristics of
PLEURAL EFFUSION

Pleural effusion is usually asymptomatic. Pleuritic chest pain usually occurs when there is significant associated inflammation, and disappears when the fluid accumulation becomes large enough to separate the inflamed layers of the pleura. With large effusions, dyspnea may occur because of disturbance of the mechanics of respiration, with a dry cough at times apparently on a reflex basis.

PHYSICAL SIGNS

Inspection

- Trachea may be shifted away from effusion.
- Chest may be enlarged, interspaces widened, in children and young adults.
- Chest expansion may be limited on affected side.

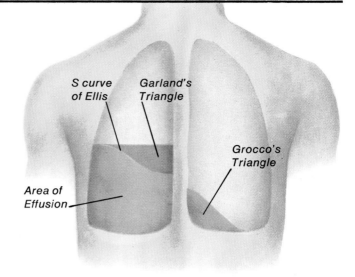

PERCUSSION IN PLEURAL EFFUSION (POSTERIOR)

Percussion

- Normal change in lower border of lung resonance with respiration (Litten's sign) is lost.
- Dullness at upper edge of fluid — level curved higher in axilla than medially (S curve of Ellis).
- Flatness over area of effusion, higher in axilla, more so when patient lies on affected side.
- Skodaic resonance, approaching tympany, in medial area above dullness (Garland's triangle).
- Dullness above diaphragm next to spine *on unaffected side* (Grocco's triangle).
- Lower border of hepatic dullness drops *with effusion on right*. Stomach tympany may be partly obscured *with effusion on left*.

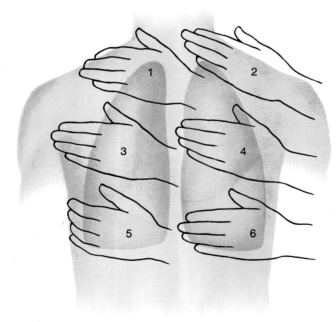

Vocal fremitus will be decreased at position 4, absent at position 6

Palpation

- Vocal fremitus decreased or absent over effusion.
- Cardiac PMI often shifted away from effusion.
- Cardiac impulses transmitted through fluid at times — interspaces are then pulsatile.

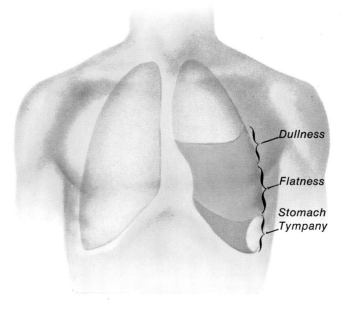

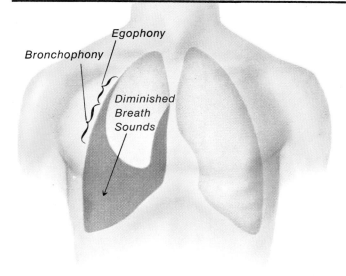

AUSCULTATORY CHANGES IN PLEURAL EFFUSION

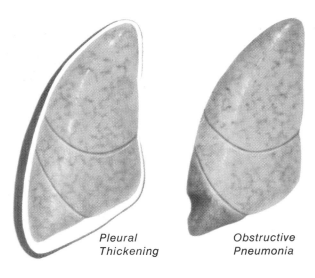

Auscultation

- Breath sounds usually normal in quality but diminished; may be absent. May be bronchial if lung is compressed; loud if lung is markedly compressed.
- Bronchophony (clear transmission of spoken voice) often heard over effusion.
- Egophony (clear transmission of spoken voice with bleating quality) usually just above effusion.

DIFFERENTIAL DIAGNOSIS

Physical findings with *a large thickened pleura* are similar to pleural effusion. The trachea will shift toward the affected side. If x-ray is inconclusive, thoracentesis will be required to differentiate. Physical findings with *consolidation secondary to obstruction* are also similar to pleural effusion. However, the trachea will not be displaced. X-ray will usually differentiate. If dullness is in the upper lobe, effusion is obviously excluded.

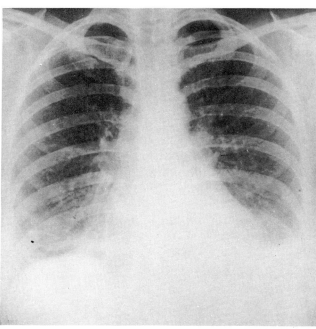

TYPICAL X-RAY OF PLEURAL EFFUSION

RADIOGRAPHIC FINDINGS

- Infrapulmonary effusion may be reflected by a high diaphragm only.
- Level of density graded, greater below than above. Upper border curved, higher laterally than medially.
- Mediastinal contents shifted to unaffected side.

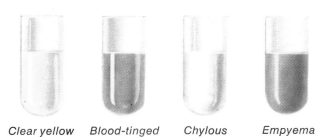

CHARACTER OF FLUID

Transudates

- Specific gravity *less than* 1.015; protein content *less than* 3 Gm/100 ml. May have high protein count in myxedema.
- Clear yellow; can be blood-tinged; few cells.

Exudates

- Specific gravity *above* 1.015; protein content *more than* 3 Gm/100 ml.
- Clear yellow with increased cells (lymphocytes, polymorphonuclear leukocytes or eosinophils); cloudy; blood-tinged; chylous; or pseudochylous.
- Chylous reflects actual increased fat; pseudochylous has large amounts of cholesterol without fat droplets.

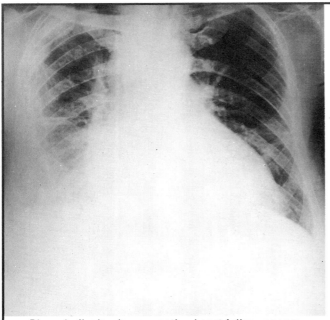
Pleural effusion in congestive heart failure accompanied by cardiomegaly and pulmonary congestion

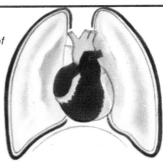
Right heart failure with engorgement of parietal pleura without effusion

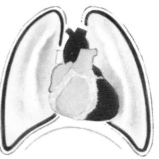
Left heart failure with engorgement of the visceral pleura without effusion

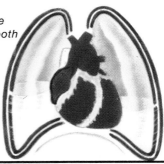
Combined heart failure with engorgement of both pleurae and effusion

Pleural effusion due to CONGESTIVE HEART FAILURE

Pleural effusion is a prominent feature of heart failure only when *both* right and left ventricular failure occurs, causing elevated pressure in vessels of the parietal pleura as well as in the visceral pleura. Diagnosis is usually not difficult, but *overdiagnosis* (attributing effusion to heart failure when it is actually due to other causes) can be an important problem.

- Onset usually gradual but can occur within hours of heart failure.
- Regresses with other edema; may persist longer.
- Usually more marked on right, sometimes only on right. If only on left, right pleural symphysis can be presumed.
- With partial pleural symphysis, transudation can cause localized effusion — phantom tumors.
- Transudate — clear yellow; does not clot. May be blood-tinged with complicating pulmonary infarct.
- May have few leukocytes.
- Typical physical findings of pleural effusion.

ASSOCIATED FINDINGS

- Fever slight. If significant, pulmonary infarct or bronchopneumonia should be considered.
- May be preceded by dyspnea at rest, orthopnea and/or paroxysmal nocturnal dyspnea.
- Nonproductive cough common.
- Bilateral moist rales, predominantly at bases, usually heard above effusion.
- Pulmonic component of second heart sound loud.
- Pulsus alternans, tachycardia, cardiomegaly on palpation, apical gallop rhythm — *with left ventricular failure.*
- Apical diastolic rumble, often opening snap of mitral valve, xiphoid presystolic gallop at times — *with mitral stenosis.*
- Distended neck veins, hepatomegaly, dependent edema — *with right heart failure.*
- X-ray will show cardiomegaly and pulmonary vascular congestion as well as effusion.

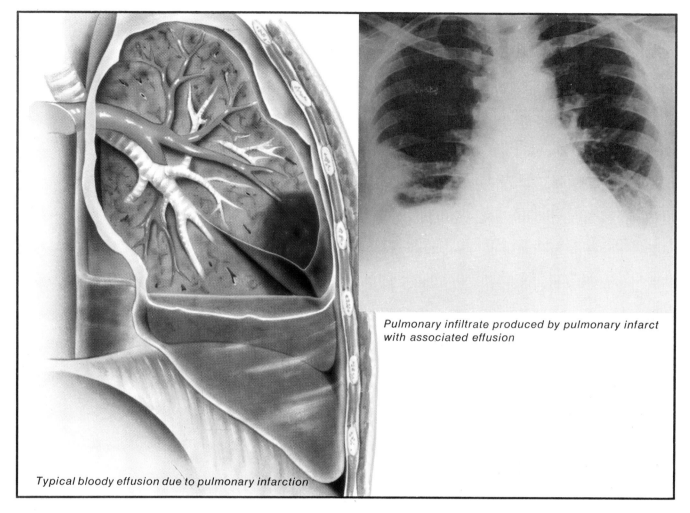

Pulmonary infiltrate produced by pulmonary infarct with associated effusion

Typical bloody effusion due to pulmonary infarction

Pleural effusion due to PULMONARY INFARCTION

The pleural effusion occurring in pulmonary infarction is usually small or moderate in quantity, accompanied by clinical signs of pleurisy. When pulmonary infarction occurs in congestive heart failure, a dual mechanism of effusion is often involved and quantity of fluid may be large.

- Occurs days after other signs and symptoms.
- Regresses gradually after a week.
- Localized to side of infarction.
- Exudate — clear yellow or blood-tinged. May have borderline specific gravity and protein content more like transudate when congestive heart failure is present.
- Few leukocytes; increased eosinophils at times.
- Typical physical findings of pleural effusion.

ASSOCIATED FINDINGS

- Fever most consistent symptom, but not always present.
- Dyspnea with shallow breathing.
- Sharp and jabbing chest pain on breathing over infarct or supraclavicular fossa.
- Nonproductive cough. Hemoptysis common; at first bright red, later dark.
- Pleural friction rub often; inspiratory moist rales more common in area of infarct.
- Signs of consolidation or pulmonary hypertension uncommon.
- ECG may show shift of QRS axis to right, right bundle branch block or atrial fibrillation.
- SGOT normal except with acute right ventricular failure and hepatic congestion.
- Serum bilirubin often rises after 3-4 days.
- Serum LDH often begins to rise within 24 hours, peaks in 2 or 3 days, may last 10 days.
- X-ray may be abnormal in 12-24 hours; with wedge-shaped or patchy infiltrate as well as effusion.
- Isotopic scan may show blocked circulation to affected part even though x-ray shows no infiltrate.

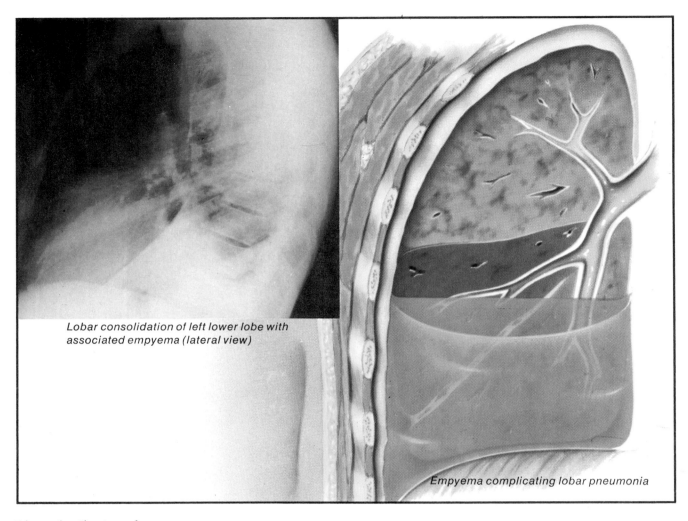

Lobar consolidation of left lower lobe with associated empyema (lateral view)

Empyema complicating lobar pneumonia

Pleural effusion due to
BACTERIAL PNEUMONIA

Pleural effusion is a common complication of bacterial pneumonias, especially the streptococcal and pneumococcal pneumonias, in which it can progress to empyema. Pleural effusion can also occur at times in viral pneumonias, fungal infections (particularly coccidioidomycosis) and Loeffler's syndrome.

- Can occur on first day of symptoms.
- Regresses within days with treatment of condition; lasts weeks to months with empyema.
- Localized to side of pneumonia.
- Exudate — clear yellow; becomes cloudy or milky with empyema.
- Increased polymorphonuclear leukocytes; bacteria not present on smear unless empyema develops.

ASSOCIATED FINDINGS

- Fever and prostration often follow severe chills.
- Headache, delirium, in severely ill.
- Skin warm and flushed.
- Severe pleuritic pain over involved lobe with tenderness and hyperesthesia of chest wall.
- Nonproductive cough, initially slight; becomes severe and productive of purulent sputum.
- Severe dyspnea with rapid, shallow breathing and expiratory grunting.
- Cyanosis at times.
- Pleural friction rub may be heard at upper border of fluid.
- Consolidation and moist rales usually obscured by fluid; may be heard if effusion is small.
- Herpetic blisters around mouth in some.
- Elevated WBC; increased polys.
- Sputum will show causative organism; often cultured from sputum, blood and pleural fluid.
- X-ray shows effusion with consolidated lobe often obscured.

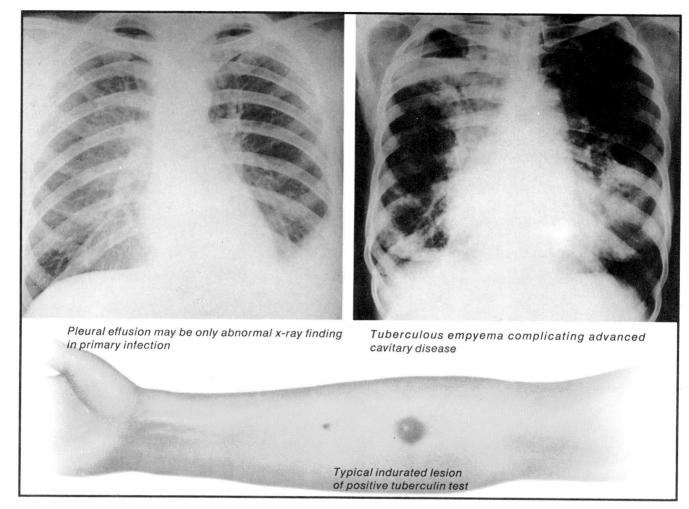

Pleural effusion may be only abnormal x-ray finding in primary infection

Tuberculous empyema complicating advanced cavitary disease

Typical indurated lesion of positive tuberculin test

Pleural effusion due to TUBERCULOSIS

Pleural effusion can occur as an isolated finding in primary tuberculous infection of children and young adults. This type of infection often regresses, only to be followed months or years later by severe tuberculosis. Effusion can also be a complication of advanced adult-type pulmonary tuberculosis, frequently followed by empyema and/or bronchopleural fistula.

- Usually an initial finding in primary infection; can occur at any stage of advanced infection.
- Often lasts weeks; may last months with empyema.
- Usually unilateral.
- Exudate — clear yellow or blood tinged with primary infection; may be cloudy with advanced infection; opaque gray with chronic empyema; silky sheen with increased cholesterol.
- Few organisms; increased lymphocytes; culture often negative in primary infection, usually positive with advanced infection.
- Typical granulomata of pleura often seen on open biopsy; at times on punch biopsy in primary infection.
- Typical physical findings of pleural effusion.

ASSOCIATED FINDINGS

- Fever and night sweats common; fever can be high without acute toxicity.
- Cough usual; initially dry and hacking, later productive of mucopurulent sputum; may be absent in primary infection.
- Pleuritic chest pain often present at onset.
- Malaise, weakness, easy fatigability, weight loss are common.
- Often fine and medium moist rales, signs of consolidation, cavitation — with advanced infection.
- Smear and culture of sputum or culture of gastric contents diagnostic with advanced infection; sputum usually negative, gastric culture at times positive, in primary infection.
- X-ray shows only effusion, sometimes hilar nodes, with primary infection; also patchy infiltrates and/or cavitation with advanced lesions.
- Tuberculin test almost always positive.

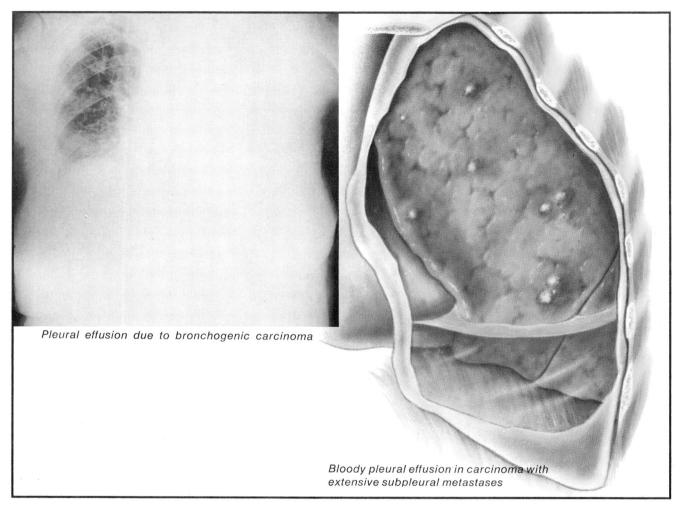

Pleural effusion due to bronchogenic carcinoma

Bloody pleural effusion in carcinoma with extensive subpleural metastases

Pleural effusion due to
MALIGNANCY

Pleural effusion can complicate many malignancies. For example, it is seen in about half of all cases of bronchogenic carcinoma, and is also frequently seen with subpleural and pleural metastatic carcinoma from many sites. Less commonly, it occurs with primary mesothelioma of the pleura, as is found in patients with asbestosis from industrial exposure.

- Onset gradual; persists during disease, recurs after chest tap.
- Unilateral; may be bilateral with metastases.
- Exudate — clear yellow or blood tinged, chylous if metastases obstruct thoracic duct.
- Red cells usual; increased lymphocytes or eosinophils at times; single or clumped cancer cells.
- LDH often high in exudate; reflects disseminated disease if also high in blood.
- Sugar level may be lower in exudate than in blood.
- Typical physical findings of pleural effusion.

ASSOCIATED FINDINGS

- Fever with associated obstructive pneumonia, lung abscess or terminal mesothelioma.
- Cough usual; change in character of any pre-existing cough; hemoptysis common.
- Weakness and malaise early symptoms; weight loss, cachexia, anemia in later stages.
- Pleuritic or nonspecific chest pain frequent.
- Localized wheeze and late onset of inspiration at times; hoarseness, chest lag, enlarged supraclavicular nodes in some.
- Clubbing of fingers, pulmonary osteoarthropathy, peripheral neuropathy, thrombophlebitis at times.
- Tumor cells sometimes identifiable in sputum.
- Survival after onset of effusion brief except in mesothelioma, where survival may be years.
- X-ray may show pulmonary densities or hilar accentuation as well as pleural effusion; major tumor mass of mesothelioma can be silhouetted on x-ray by artificial pneumothorax.
- Bronchoscopy diagnostic in about 50% of cases.

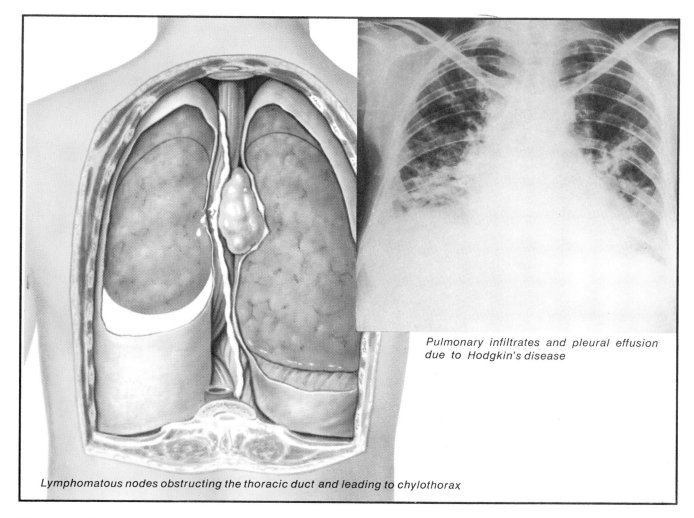

Pulmonary infiltrates and pleural effusion due to Hodgkin's disease

Lymphomatous nodes obstructing the thoracic duct and leading to chylothorax

Pleural effusion due to LYMPHOMA

Pleural effusion and/or pulmonary infiltration may occur in about 20% of patients with lymphomas such as Hodgkin's disease, lymphosarcoma, reticulum cell sarcoma and giant follicular lymphoma. It is usually an indication of Stage IV progression, since it reflects extranodal involvement. A chylous pleural effusion, or chylothorax, however, does not necessarily represent Stage IV progression, as it is sometimes produced by nodal obstruction of the thoracic duct.

- Onset is gradual and insidious; may regress spontaneously or with chemotherapy.
- Usually exudate — clear yellow, may be chylous or pseudochylous; may be transudate in Hodgkin's disease.
- Lymphomatous cells at times; increased eosinophils in Hodgkin's disease.
- Typical physical findings of pleural effusion.

ASSOCIATED FINDINGS

- Fever, anorexia, night sweats, weight loss and weakness with systemic involvement.
- Usually no pleuritic pain; pain in lymphomatous nodes for 30 to 60 minutes after ingestion of alcohol common in Hodgkin's disease.
- Pruritus common; skin may be hyperpigmented in Hodgkin's disease.
- Herpes zoster in skin may be early sign.
- Enlarged lymph nodes, palpable spleen, liver and abdominal masses.
- Moist rales with underlying pulmonary disease.
- Anemia at times.
- Uric acid often elevated; abnormal SGOT, alkaline phosphates and BSP with liver involvement.
- Serum electrophoresis shows increased alpha 1 and alpha 2 globulin.
- Gamma globulin may be increased or decreased.
- Often anergy to tuberculin and mumps antigen.
- X-ray usually shows pulmonary infiltration (miliary or nodular, at times with cavitation) as well as pleural effusion.
- Definitive diagnosis by node biopsy.

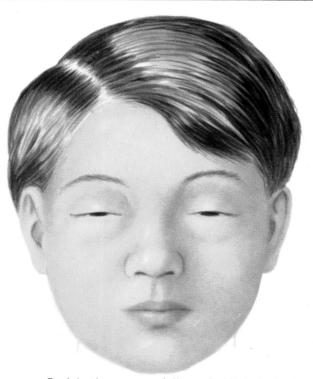

Facial edema, especially periorbital, part of the generalized edema of the nephrotic syndrome

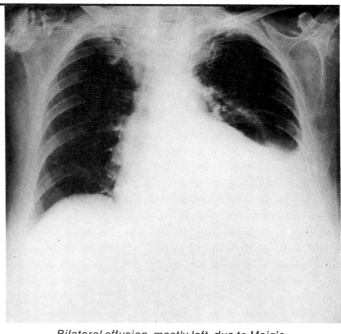

Bilateral effusion, mostly left, due to Meig's syndrome.

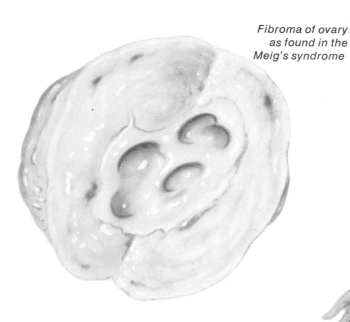

Fibroma of ovary as found in the Meig's syndrome

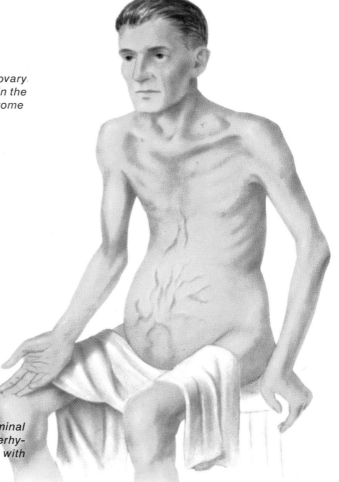

Wasting, jaundice, ascites, dilated abdominal veins, spider angiomata and palmar erhythema are found in cirrhotic patients with pleural effusion

Pleural effusion due to CIRRHOSIS OF LIVER

Pleural effusion can occur with any type of cirrhosis — alcoholic, portal, postnecrotic or schistosomal. It is always associated with ascites. Diagnosis should only be made following aspiration since effusion in cirrhotic patients may reflect complicating tuberculosis or pulmonary infarction.

- Onset usually gradual; can be abrupt.
- May come and go with changes in treatment, salt intake and status of liver.
- Usually on right, may be bilateral.
- Transudate — clear yellow; darker with jaundice.
- Few cells.
- Typical physical findings on pleural effusion.

ASSOCIATED FINDINGS

- Dyspepsia, fullness and bloating with onset of ascites.
- Pleuritic pain not present.
- Usually menstrual irregularities in female; decreased libido, impotence, gynecomastia and loss of axillary hair at times in male.
- Wasting of muscle and subcutaneous tissue.
- Jaundice, vascular spiders in skin, palmar erythema common.
- Flared ribs and enlarged collateral veins over abdomen accompany ascites.
- Liver may be palpable; nodularity suggests postnecrotic cirrhosis; spleen often enlarged.
- SGOT and SGPT elevated; cephalin-cholesterol flocculation abnormal; bromsulphalein excretion reduced; serum albumin often low.
- X-ray shows high diaphragm as well as pleural effusion.

Pleural effusion due to MEIG'S SYNDROME

Pleural effusion is an unusual complication of ovarian tumors (usually benign fibromas) in middle-aged women. When the effusion is in the right pleural cavity and accompanied by ascites, the condition is termed Meig's syndrome. The mechanism of effusion is unknown.

- Onset gradual.
- Regresses after removal of tumor.
- Unilateral on right side; rarely bilateral.
- Transudate — clear yellow.
- Few cells; no tumor cells.
- Typical physical findings of pleural effusion.

ASSOCIATED FINDINGS

- No pleuritic pain.
- Marked ascites always present.
- Tumor can be felt by bimanual examination.

Pleural effusion due to NEPHROTIC SYNDROME

Pleural effusion is an aspect of the generalized edema of the nephrotic syndrome. It is a result of the hypoproteinemia and tendency to salt and water retention which are characteristic of the disease. The effusion is seldom associated with congestive heart failure, as heart failure is uncommon in the nephrotic syndrome.

- Onset gradual and insidious.
- Lasts for months if condition is untreated.
- Recurs after chest tap.
- Unilateral or bilateral.
- Transudate — clear yellow.
- Few cells.
- Typical physical findings of pleural effusion.

ASSOCIATED FINDINGS

- Pleuritic pain not present.
- Generalized edema and ascites common; scrotal or vulvar edema may be marked.
- Pallor frequently severe, suggesting lower hemoglobin than is actually present.
- Red urine with glomerulonephritis.
- Dilatation of veins of lower thorax and abdomen.
- Hemoglobin sometimes elevated; WBC normal.
- Serum albumin low, serum complement low at times; total serum lipids increased.
- Proteinuria always marked; hematuria, pyuria and cylindruria at times; doubly refractile bodies by polarized light.
- X-ray usually shows small heart as well as pleural effusion.

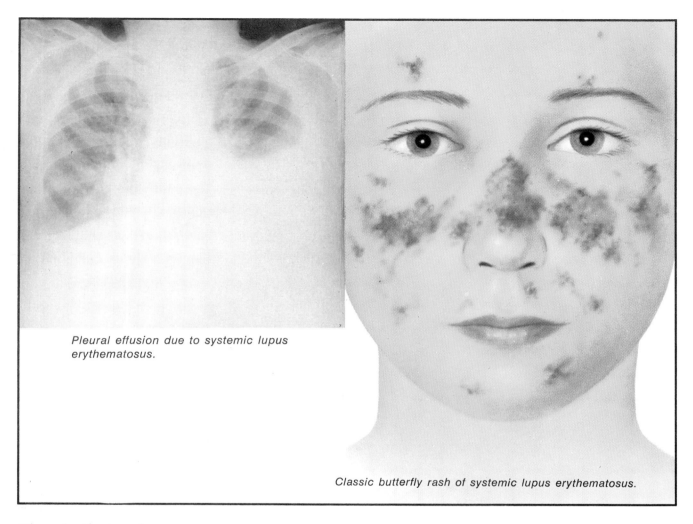

Pleural effusion due to systemic lupus erythematosus.

Classic butterfly rash of systemic lupus erythematosus.

Pleural effusion due to SYSTEMIC LUPUS ERYTHEMATOSUS

Pleural effusion can occur in several connective tissue diseases. It is particularly prominent in systemic lupus erythematosus, where it may be a most important early diagnostic sign. Systemic lupus erythematosus is predominantly seen in women of childbearing age.

- Onset sudden or gradual; lasts days to weeks; may regress spontaneously.
- Usually small and unilateral.
- Does not tend to recur after chest tap.
- Exudate — clear yellow.
- Increased polymorphonuclear cells.
- Typical physical findings of pleural effusion.

ASSOCIATED FINDINGS

- Fever most common symptom; recurrent; usually accompanies effusion.
- Fatigue, arthritis or arthralgia, malaise and weight loss frequent.
- Erythematous and edematous rash in butterfly distribution over face; also may be at mucocutaneous junctions, arms, knees, elbows, palms and soles; may heal with atrophy.
- Alopecia often associated with rash.
- Signs of pericarditis frequent.
- Friction rub and moist rales with associated pneumonia.
- Lymphadenopathy, enlarged liver and spleen, tender abdomen common.
- Proteinuria, pyuria, cylinduria and high BUN.
- Leukopenia; increased mononuclear cells; anemia common; ESR and CRP elevated.
- Antinuclear antibody usually present; often lupus cells after incubation of serum.
- Gamma globulin increased; biologic false positive serologic test for syphilis frequent.
- X-ray may show pneumonic infiltrate as well as effusion.
- Syndrome occasionally drug precipitated.

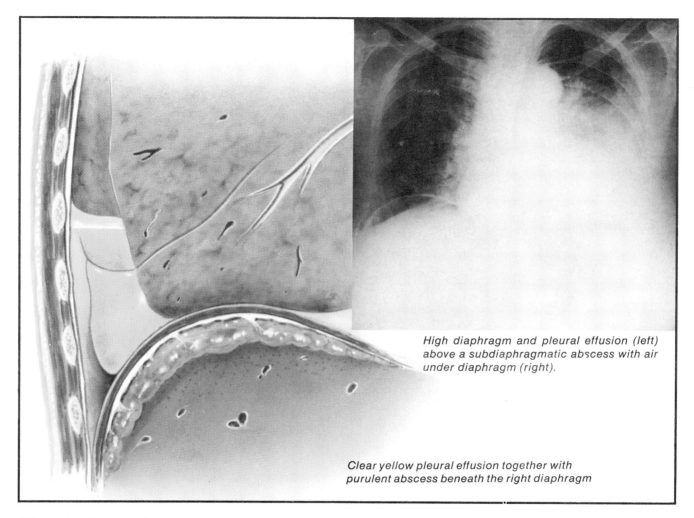

High diaphragm and pleural effusion (left) above a subdiaphragmatic abscess with air under diaphragm (right).

Clear yellow pleural effusion together with purulent abscess beneath the right diaphragm

Pleural effusion due to
SUBDIAPHRAGMATIC OR HEPATIC ABSCESS

Pleural effusion, along with an elevated diaphragm producing pain and splinting, can complicate a subdiaphragmatic or hepatic abscess. It can occur in a noninfected pleural space as a reaction to adjacent inflammation and infection. The infection can also extend through a ruptured diaphragm and produce an infected pleural effusion.

- Onset usually gradual, may be abrupt; lasts weeks or months unless treated.
- Usually unilateral, most often on right side.
- Exudate — usually clear yellow; may be opaque with empyema; chocolate colored with amebiasis.
- Increased polys, bacteria and debris with empyema.
- Aspirating needle may enter pus above or below diaphragm — on inspiration, outer edge of needle rises when above, pus runs out faster when below.
- Typical physical findings of pleural effusion.

ASSOCIATED FINDINGS

- Recurrent fever.
- Right upper quadrant pain, at times pleuritic.
- Pain usually referred to right shoulder: anteriorly with abscess below front of diaphragm; to point of shoulder with abscess near mid-diaphragm; posteriorly with abscess below back of diaphragm.
- Productive cough with purulent sputum at times.
- Abscess and effusion may occur together with inciting infection — phlebitis or ascending cholangitis; may also occur after long silent period.
- Friction sounds over liver with subdiaphragmatic abscess.
- Tenderness and/or bulging of right lobe of liver.
- Jaundice when cholangitis is inciting cause.
- WBC elevated, 20-30,000 in bacterial abscess.
- Complement fixation positive, *Entamoeba hystolytica* often found in stool with amebiasis.
- X-ray shows tenting of diaphragm as well as small to moderate effusion; distortion of liver, or abscess itself, best visualized by radioisotopic scan.

DIFFERENTIAL DIAGNOSIS OF PLEURAL EFFUSION

	Onset	Location	Character of Fluid
CONGESTIVE HEART FAILURE	Gradual, within hours of other symptoms	Usually on right, may be only on right	Transudate — usually clear yellow; does not clot; may have few leukocytes
PULMONARY INFARCTION	Within days of other findings	Localized to side of infarction	Exudate — clear yellow or blood-tinged; may have borderline specific gravity; few leukocytes; increased eosinophils at times
BACTERIAL PNEUMONIA	May be on first day of illness	Localized to side of pneumonia	Exudate — clear yellow, becomes cloudy or milky with empyema; increased polys; bacteria on smear in empyema
TUBERCULOSIS	Initial sign in primary, at any stage in advanced	Usually unilateral	Exudate — clear yellow or blood-tinged in primary, cloudy or opaque in advanced; increased lymphocytes; few organisms; culture often negative in primary
MALIGNANCY	Gradual	Usually unilateral, may be bilateral with metastases	Exudate — clear yellow or blood-tinged, chylous with thoracic duct obstruction; red cells; increased lymphocytes; eosinophils at times; single or clumped cancer cells

Duration	Associated Symptoms and Signs	Key Laboratory Data
May regress with other edema; may persist	Dyspnea; orthopnea; moist rales; nonproductive cough; tachycardia; gallop rhythm; loud P_2; venous distention; dependent edema; hepatomegaly; pulsus alternans; mitral diastolic rumble	Cardiomegaly, pulmonary vascular congestion as well as effusion on x-ray
Regresses after a week	Fever; pleuritic chest pain; dyspnea; tachycardia; patch of inspiratory moist rales; pleural friction rub; signs of consolidation; nonproductive cough; hemoptysis at times	ECG shows shift of QRS axis to right at times; infiltrate as well as effusion on x-rays; SGOT usually normal; serum LDH and bilirubin often elevated
Few days with therapy; months with empyema	Fever; prostration; headache; pleuritic pain; dyspnea; productive cough; cyanosis at times; pleural friction rub; signs of consolidation and moist rales obscured; herpetic blisters around mouth	Elevated WBC count; increased polys; organism on sputum smear; can be cultured from sputum, blood or pleural fluid; consolidated lobe often obscured by effusion on x-ray
Usually weeks; with empyema may last months	Fever; night sweats; weakness; weight loss; fine and medium moist rales mostly over upper lung; cavernous breath sounds over cavities; cough often absent in primary, productive in advanced cases	Tuberculin test and gastric culture positive; sputum culture negative in primary, positive in advanced cases; hilar nodes as well as effusion on x-ray in primary, extensive changes with effusion in advanced cases
Persists during disease; recurs after chest tap	Change in character of existing cough; hemoptysis; weakness; malaise; chest pain; weight loss; anemia; localized wheezing; enlarged nodes; moist rales; pulmonary osteoarthropathy; thrombophlebitis	Pulmonary densities or hilar accentuation as well as effusion on x-ray; tumor cells at times in sputum; diagnosis possible by bronchoscopy in 50% of cases

(Continued on next page.)

DIFFERENTIAL DIAGNOSIS OF PLEURAL EFFUSION (continued)

	Onset	Location	Character of Fluid
LYMPHOMA	Gradual and insidious	Unilateral or bilateral	Exudate — clear yellow, may be chylous or pseudochylous; may be transudate; may have increased eosinophils in Hodgkin's disease; lymphomatous cells
CIRRHOSIS OF LIVER	Usually gradual, may be abrupt	Usually on right, may be bilateral	Transudate — clear yellow, darker with jaundice; few cells
MEIG'S SYNDROME	Gradual	Unilateral right, rarely bilateral	Transudate — clear yellow; few cells; no tumor cells
NEPHROTIC SYNDROME	Gradual and insidious	Unilateral or bilateral	Transudate — clear yellow; few cells
SYSTEMIC LUPUS ERYTHEMATOSUS	Gradual or sudden	Usually unilateral	Exudate — clear yellow; increased polys
SUBDIAPHRAGMATIC OR HEPATIC ABSCESS	Usually gradual, may be abrupt	Usually unilateral, mostly at right	Exudate — clear yellow, may be opaque with empyema, chocolate-colored with amebiasis; increased polys; debris with empyema

Duration	Associated Symptoms and Signs	Key Laboratory Data
May regress with or without chemotherapy	Fever; anorexia; weight loss; pruritus; pain in involved areas after ingestion of alcohol and hyperpigmented skin in Hodgkin's disease; herpes zoster; enlarged lymph nodes; palpable liver and spleen; moist rales; anemia at times	Serum uric acid elevated; anergy to tuberculin and mumps antigen common; infiltrate as well as effusion on x-ray; diagnosis by node biopsy
Comes and goes with therapy and diet changes	Dyspepsia; fullness; bloating; jaundice; wasting; loss of axillary hair; ascites; distended abdominal veins; vascular spiders; palmar erythema; hepatic fetor to breath; palpable liver and spleen	SGOT and SGPT elevated; cephalin-cholesterol flocculation abnormal; BSP excretion reduced; high diaphragm as well as effusion on x-ray
Regresses with removal of tumor	Marked ascites; tumor can be felt by bimanual examination; pleuritic pain absent	High diaphragm as well as effusion on x-ray
Months if untreated	Red urine; generalized edema; often facial; scrotal and vulvar edema may be marked; ascites; dilatation of veins of lower thorax and abdomen; pallor	Proteinuria, pyuria, cylindruria and hematuria; doubly refractile bodies on polarized light; serum albumin low, serum complement low at times; total serum lipids increased; small heart as well as effusion on x-ray
Days to weeks; may regress abruptly	Fever; malaise; arthritis or arthralgia; chest pain; rash; alopecia; pleural friction rub; moist rales; lymphadenopathy; enlarged liver and spleen; tender abdomen	Antinuclear antibody; lupus cells often in incubated serum; anemia; leukopenia; gamma globulin increased; proteinuria, pyuria, cylindruria, and high BUN; infiltrate as well as effusion on x-ray; false positive test for syphillis frequent
Weeks to months or until treated	Fever; right upper quadrant pain referred to shoulder; productive cough; friction sounds over liver; splinting of right side of chest; tenderness of right lobe of liver; jaundice	WBC high in bacterial abscess; complement fixation positive and *Entamoeba histolytica* in stool with amebiasis; tenting of diaphragm and small to moderate effusion on x-ray; abscess seen on liver scan

GALLOP RHYTHM

Gallop rhythm is one of the most common and diagnostically important physical findings in heart disease.

While occurring at times without heart disease — in such "high-output" states as fever, anemia and hyperthyroidism — gallop rhythm in the adult patient usually reflects an excessive pressure or volume load on the heart. Thus, gallop rhythm may be important evidence that the heart disease present is of dynamic significance. Further, the location and timing of the gallop in conjunction with other findings may serve as helpful clues in establishing the specific cardiac diagnosis.

The general characteristics of gallop rhythm and ten important causes of gallop rhythm are described and then compared in tabular form.

GENERAL PRINCIPLES OF GALLOP RHYTHM

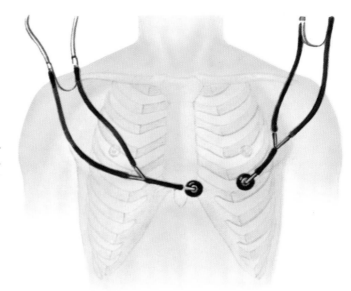

Gallop rhythm refers to the auscultatory effects produced by an audible *three-sound* sequence to each heart beat, rather than the normal *two-sound* (lub-dub) pattern of the first and second heart sounds.

The abnormal extra sound never occurs in systole, but is always a diastolic event. It is brief and of low frequency like the first and second heart sounds. Thus, triple patterns produced by murmur or by opening snap are *not* gallop rhythm.

Gallop rhythms produced in the left heart are heard best at the cardiac apex; those produced in the right heart are heard best in the xiphoid region and along the lower left sternal border.

When the abnormal extra sound occurs with pulses of 100 or below, the resemblance between the rhythm and a horse's gallop is usually not so pronounced.

Stethoscopes indicate typical locations of maximal intensity of right and left ventricular gallop sounds.

PROTODIASTOLIC GALLOP RHYTHM

Produced by an accentuation of the *third heart sound* in early diastole. Can be mimicked by repeating the word "Kentucky."

The normal third sound is produced in the open left ventricle and atrium toward the end of the period of rapid emptying of the atrial contents into the ventricle. The sound is usually inaudible in normal adults but may become exaggerated and audible in abnormal states when the volume of early filling is increased, or when there is incomplete emptying during the prior systole.

- Usually closer to second heart sound than to first heart sound.

- Occurs between .12 and .18 second after aortic valve closure (A_2).
- Heard mostly on left side; usually maximal at cardiac apex, at times associated with a palpable thrust.
- Heard best when patient recumbent.
- Frequent in aortic or mitral insufficiency and in heart failure.
- May be reduced by tourniquets to extremities.
- May occur in right heart; typically maximal along lower left sternal border, may be increased by inspiration.

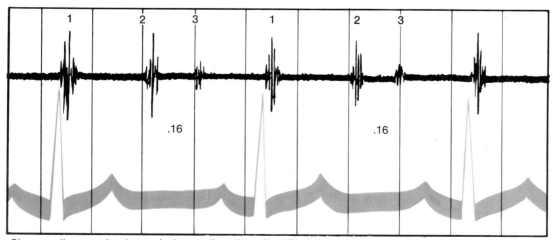

Phonocardiogram showing typical protodiastolic gallop (3). It follows the second sound (2) by .12 to .20 second.

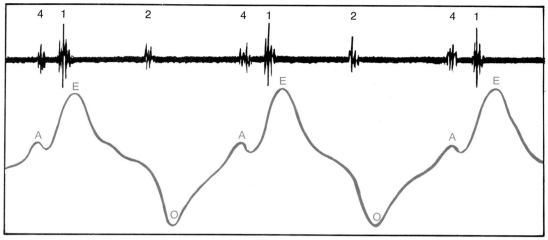

Phonocardiogram showing presystolic gallop (4) and its relation to first sound (1) — above. Below, an apex cardiogram shows the occurrence of the a wave of atrial contraction together with the presystolic gallop sound.

PRESYSTOLIC GALLOP RHYTHM

Produced by an accentuation of the *fourth heart sound* in late diastole. Can be mimicked by repeating the word "Tennessee."

The normal fourth heart sound is produced by vibration of the blood column in the open atrium and ventricle during atrial contraction. Although usually inaudible in normal adults, the sound may become exaggerated when the open atrium and ventricle are already overfilled.

- Usually quite close to first heart sound.
- Occurs about .10 second after P wave.
- Does not occur when atrium is markedly distended and contracts poorly, as in mitral insufficiency; does not occur during atrial fibrillation.

DOUBLE GALLOP SOUNDS

When both protodiastolic and presystolic gallop sounds are audible with a slower heart rate and relatively short PR, there is a resultant *quadruple rhythm*. When both sounds are present with a more rapid heart rate or longer PR, there is a fusing of the abnormal sounds in mid-diastole. The result is a triple rhythm with a very loud gallop sound — *a summation gallop*.

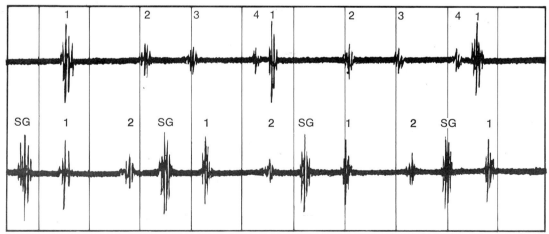

The phonocardiographic strip above shows separate presystolic and protodiastolic gallop; the strip below their fusion — resulting in a summation gallop.

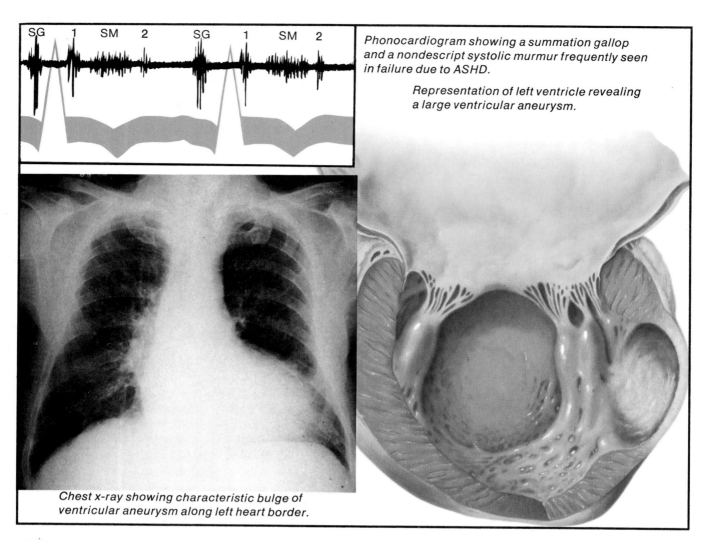

Phonocardiogram showing a summation gallop and a nondescript systolic murmur frequently seen in failure due to ASHD.

Representation of left ventricle revealing a large ventricular aneurysm.

Chest x-ray showing characteristic bulge of ventricular aneurysm along left heart border.

Gallop rhythm in
ARTERIOSCLEROTIC HEART DISEASE

Gallop rhythm is a common occurrence during the course of arteriosclerotic heart disease. It is often present in acute pulmonary edema, myocardial infarction and angina pectoris, and frequently persists in heart failure and in the presence of ventricular aneurysm.

- Heard best at cardiac apex.
- May be presystolic only — in incipient or subsiding heart failure, during anginal episodes or in the presence of ventricular aneurysm without heart failure; may be protodiastolic alone at times — in heart failure with atrial fibrillation or weak atrial contraction.
- Double apical impulse in ventricular aneurysm may be due to associated "atrial kick," the palpable counterpart of presystolic gallop.
- Both gallop sounds may occur in heart failure with slower heart rate; may be a loud summation gallop with more rapid heart rate.

ASSOCIATED FINDINGS

- Dyspnea, orthopnea, paroxysmal nocturnal dyspnea, nonproductive cough in heart failure.
- Pressing substernal chest pain in angina pectoris or infarction.
- May be asymptomatic in ventricular aneurysm.
- Peripheral cyanosis at times.
- Blood pressure may be normal or low.
- Pulsus alternans often present.
- First heart sound often diminished; paradoxical splitting of second sound at times — in severe heart failure, left bundle branch block or acute myocardial infarction.
- Transitory systolic murmur at apex at times; usually not holosystolic.
- Basilar moist inspiratory rales with left heart failure; distended neck veins, hepatomegaly and edema with associated right heart failure.
- ECG usually shows evidence of infarction.
- X-ray may show cardiomegaly, vascular congestion; fluoroscopy will show paradoxical pulsation of left ventricle when aneurysm is present.

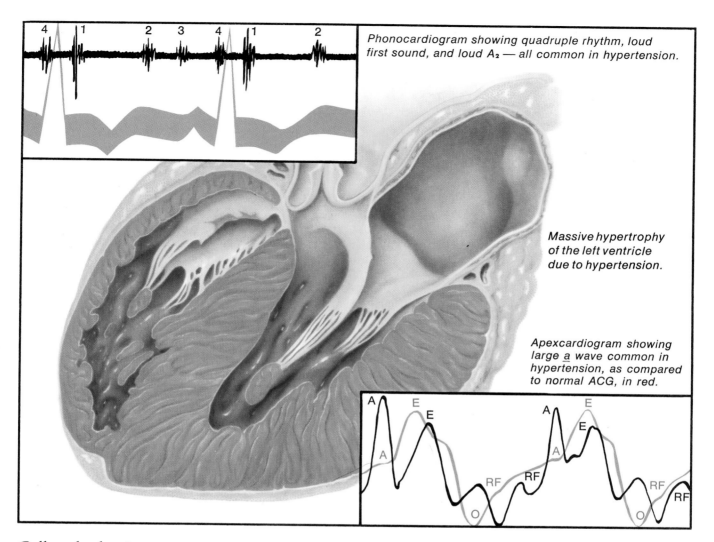

Phonocardiogram showing quadruple rhythm, loud first sound, and loud A₂ — all common in hypertension.

Massive hypertrophy of the left ventricle due to hypertension.

Apexcardiogram showing large a wave common in hypertension, as compared to normal ACG, in red.

Gallop rhythm in
HYPERTENSIVE HEART DISEASE

Gallop rhythm is a prominent feature of left heart failure in hypertensive heart disease. It can also be present in systemic arterial hypertension without overt heart failure, the high pressure of the ventricle during systole apparently altering ventricular compliance and leading to a more forceful left atrial contraction.

- Heard best at cardiac apex.
- May be presystolic only in incipient or subsiding heart failure, or without heart failure when arterial pressure is very high.
- Presystolic apical "atrial kick" may be present.
- May be only protodiastolic in heart failure with atrial fibrillation.
- Both sounds may occur in heart failure with slower heart rate; a loud summation gallop may occur with more rapid heart rate.

ASSOCIATED FINDINGS

- Dyspnea, orthopnea, paroxysmal nocturnal dyspnea, nonproductive cough in heart failure.
- Headache, blurred vision, hematuria in severe hypertension.
- Polyuria and polydipsia when hypokalemia complicates severe hypertension.
- Elevated systolic and diastolic arterial pressures.
- Forceful peripheral pulse and vigorous sustained apical impulse.
- Loud first heart sound and A_2; loud P_2 in heart failure.
- Basilar moist inspiratory rales reflect left ventricular failure.
- Distended neck veins, hepatomegaly and edema with associated right heart failure.
- Retinal fundi show arteriolar narrowing and at times hemorrhages, exudates and papilledema.
- ECG shows left ventricular hypertrophy and secondary ST-T abnormality of "systolic overload."
- X-ray shows cardiomegaly and some dilatation of aortic arch in older patients.

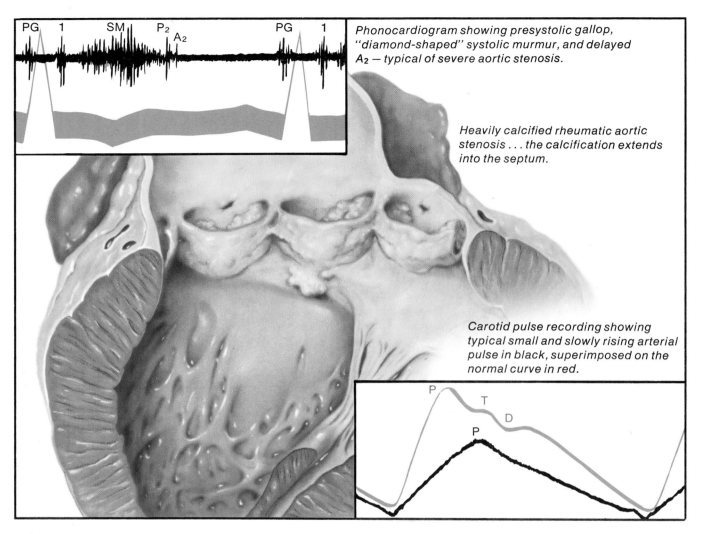

Phonocardiogram showing presystolic gallop, "diamond-shaped" systolic murmur, and delayed A_2 — typical of severe aortic stenosis.

Heavily calcified rheumatic aortic stenosis... the calcification extends into the septum.

Carotid pulse recording showing typical small and slowly rising arterial pulse in black, superimposed on the normal curve in red.

Gallop rhythm in AORTIC STENOSIS

Gallop rhythm, in addition to the characteristic murmur, is a common finding in aortic stenosis and is of considerable value in assessing the severity of obstruction. Typical aortic stenosis can occur not only in rheumatic heart disease, but may also be congenital in origin or may reflect a late calcification of a bicuspid aortic valve.

- Heard best at cardiac apex.
- May be presystolic alone with heart failure, or without heart failure if considerable valve obstruction is present.
- Double apical impulse due to "atrial kick" with severe stenosis.
- At times protodiastolic gallop may occur alone.
- Both sounds or summation gallop may occur in heart failure.

ASSOCIATED FINDINGS

- Dyspnea, orthopnea, paroxysmal nocturnal dyspnea, nonproductive cough with heart failure.
- Pressing substernal pain or syncope on exertion.
- Narrow arterial pulse pressure; arterial pulses small and slowly rising.
- Often pulsus alternans.
- Systolic thrill at 2nd right interspace, in supraclavicular notch and over carotid arteries.
- Ejection systolic murmur after first heart sound, radiating to neck and apex; ejection sound frequent after first heart sound.
- First sound normal or loud; A_2 often obscured and reduced in intensity; splitting of second sound reduced — paradoxical splitting may occur in severe cases.
- Moist rales with left heart failure; distended veins, hepatomegaly, edema with right heart failure.
- ECG shows left ventricular hypertrophy with secondary ST-T abnormality of "systolic overload" type; may show bundle branch block or complete block when calcification extends into septum.
- X-ray shows normal heart size or cardiomegaly and dilatation of ascending aorta; aortic valve calcification seen in most adults with image intensified fluoroscopy.

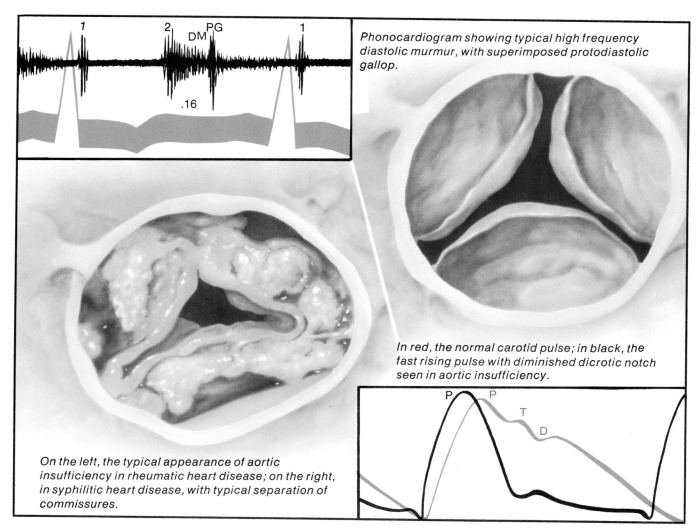

Phonocardiogram showing typical high frequency diastolic murmur, with superimposed protodiastolic gallop.

On the left, the typical appearance of aortic insufficiency in rheumatic heart disease; on the right, in syphilitic heart disease, with typical separation of commissures.

In red, the normal carotid pulse; in black, the fast rising pulse with diminished dicrotic notch seen in aortic insufficiency.

Gallop rhythm in
AORTIC INSUFFICIENCY

Gallop rhythm is frequently present in aortic insufficiency and is an important guide in assessing the severity of the regurgitation. Aortic insufficiency can occur in rheumatic or syphilitic heart disease or as a complication of Marfan's syndrome, ankylosing spondylitis or aortic dissection. It may also occur with congenital bicuspid valve when associated with coarctation, and may sometimes be seen as a result of trauma.

- Usually protodiastolic only with heart failure, or without it when due to rapid overfilling.
- Presystolic gallop uncommon even with large *a* waves in left atrium.
- Double gallop sounds or summation gallop infrequent even with marked congestive failure; when both gallop sounds do occur, additional aortic stenosis should be considered; bisferiens pulse usually present.

ASSOCIATED FINDINGS

- Awareness of heartbeat and neck pulsation at times.
- Excessive sweating and heat intolerance.
- Angina pectoris-like pain at times on exertion; exertional dyspnea, orthopnea, paroxysmal nocturnal dyspnea in heart failure.
- Visible and palpable arterial pulses, rapidly rising and falling (water-hammer pulse); dicrotic notch often lost; forceful, diffuse apical impulse.
- Normal or increased first sound, diminished with advanced disease; early systolic ejection sound.
- Decrescendo diastolic blowing murmur may last through diastole, maximal at 2nd right or 3rd and 4th left interspaces, radiating to apex.
- Sometimes diastolic rumble at apex, beginning after gallop and lasting through late diastole (Austin-Flint murmur).
- Bilateral moist rales with left ventricular failure; distended neck veins, hepatomegaly, edema with right heart failure.
- ECG: left axis deviation more common than in aortic stenosis, but high voltage QRS and typical ST-T changes of left ventricular hypertrophy less frequent.
- X-ray shows cardiomegaly with elongated boot-shaped heart, dilated aorta.

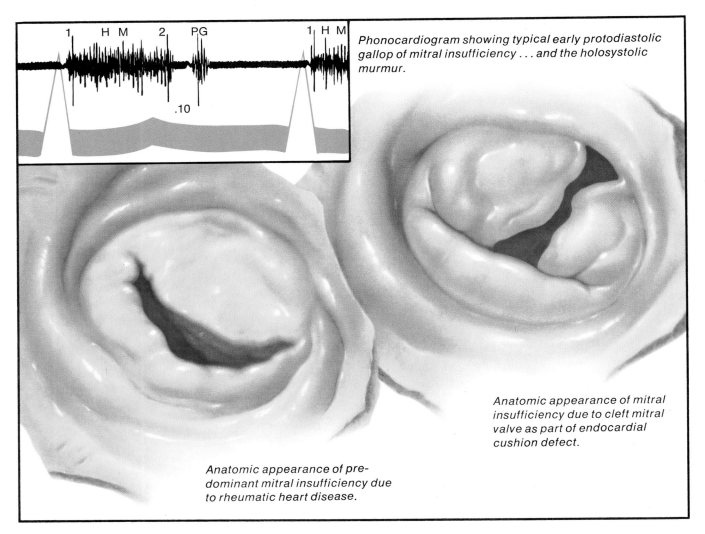

Phonocardiogram showing typical early protodiastolic gallop of mitral insufficiency... and the holosystolic murmur.

Anatomic appearance of predominant mitral insufficiency due to rheumatic heart disease.

Anatomic appearance of mitral insufficiency due to cleft mitral valve as part of endocardial cushion defect.

Gallop rhythm in
MITRAL INSUFFICIENCY

Gallop rhythm can be an important guide in assessing the dominant lesion in mitral insufficiency with associated mitral stenosis, since apical gallop rhythm does not occur when stenosis is prominent. Although insufficiency of the mitral valve may occur in arteriosclerotic heart disease from papillary muscle dysfunction or in idiopathic hypertrophic stenosis from valve displacement, it is more commonly a result of valve deformity as in rheumatic heart disease or congenital endocardial cushion defects. Less frequently, it is a complication of Marfan's or Ehlers-Danlos syndrome.

- Mainly protodiastolic, due to rapid filling in early diastole; sharp in quality, earlier than usual protodiastolic gallop (.10 to .12 second after A_2).
- Palpable early diastolic shock or "knock" with marked insufficiency.
- Presystolic gallop rare in left heart, may occur with small left atrium; heard along left sternal border at times from right heart as result of pulmonary hypertension.

ASSOCIATED FINDINGS

- Fatigue, cough, palpitation early.
- Dyspnea, orthopnea, paroxysmal nocturnal dyspnea from heart failure.
- Little water-hammer pulse with rapid rise and fall.
- Vigorous apical impulse.
- First sound normal or loud, often obscured by murmur; diminished with advanced failure; sometimes exaggerated inspiratory split P_2; first sound may be split with right ventricular failure.
- Holosystolic apical murmur, radiating to axilla; increased by vasopressors, may be reduced by amyl nitrite.
- Short diastolic rumble following gallop at times.
- Opening snap at times — mobile anterior leaflet.
- Atrial fibrillation frequent.
- ECG shows left ventricular hypertrophy; P mitrale more frequent (when rhythm is sinus).
- X-ray shows enlarged left ventricle, left atrium, at times enlarged pulmonary arteries and right ventricle.

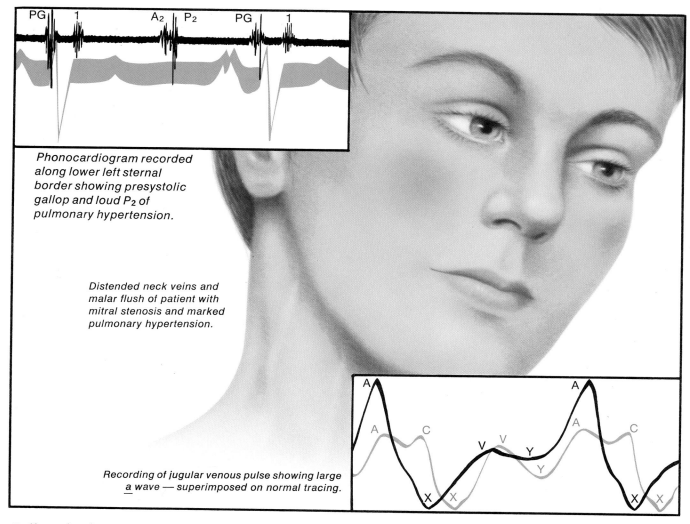

Phonocardiogram recorded along lower left sternal border showing presystolic gallop and loud P₂ of pulmonary hypertension.

Distended neck veins and malar flush of patient with mitral stenosis and marked pulmonary hypertension.

Recording of jugular venous pulse showing large *a* wave — superimposed on normal tracing.

Gallop rhythm in PULMONARY HYPERTENSION

Gallop rhythm along the left sternal border is of diagnostic importance in pulmonary disease, whether due to emphysema, fibrosis, infiltration or vascular obliteration. The gallop is evidence that the pulmonary hypertension produced by the lung disease is putting a serious strain on right ventricular function. The same gallop may occur in mitral stenosis or mitral insufficiency when there is considerable pulmonary hypertension.

- Presystolic gallop in xiphoid region or along lower left sternal border common, with or without heart failure; protodiastolic gallop alone uncommon, even with marked right-sided heart failure.
- Double and summation gallops rare on right side; combination of protodiastolic at apex with presystolic at left sternal border most common in mitral insufficiency.
- Prominent *a* wave in jugular venous pulse frequent with presystolic gallop.

ASSOCIATED FINDINGS

- Exertional or resting dyspnea from pulmonary disease; exertional dyspnea, orthopnea and paroxysmal nocturnal dyspnea in mitral valve disease.
- Cyanosis and cough common.
- Blood pressure and pulse normal or diminished.
- Right ventricular heave often palpable along left sternal border.
- First heart sound usually normal; P₂ loud, may be palpable.
- Ejection sound in early systole if pulmonary artery is dilated.
- Ejection-type systolic murmur along left sternal border; diastolic blowing murmur of pulmonic incompetence in mitral stenosis when hypertension is marked.
- Typical physical findings of primary disease.
- Distended neck veins, hepatomegaly and edema reflect right ventricular failure.
- ECG shows right ventricular hypertrophy, often atrial fibrillation.
- X-ray shows large pulmonary arteries and characteristic findings of primary disease.

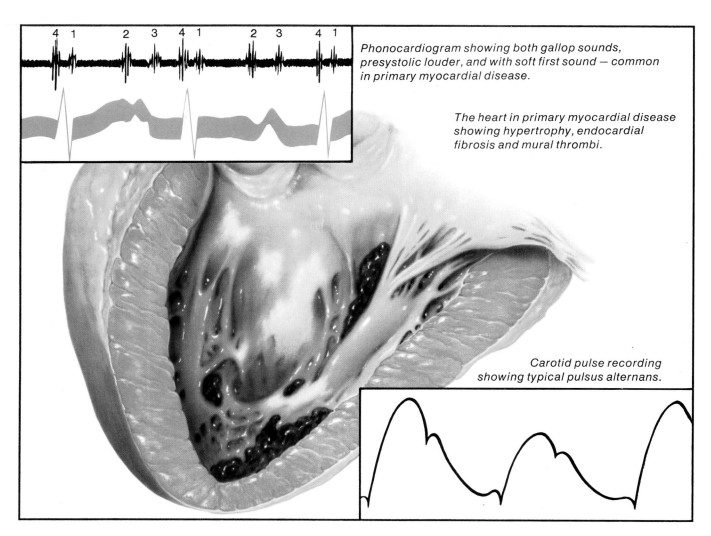

Phonocardiogram showing both gallop sounds, presystolic louder, and with soft first sound — common in primary myocardial disease.

The heart in primary myocardial disease showing hypertrophy, endocardial fibrosis and mural thrombi.

Carotid pulse recording showing typical pulsus alternans.

Gallop rhythm in PRIMARY MYOCARDIAL DISEASE

Gallop rhythm is the most characteristic physical finding in primary myocardial disease — a syndrome characterized by progressive weakness and hypertrophy of the ventricles (especially the left), recurrent heart failure, intracardiac mural thrombi and pulmonary embolization. The gallop is a critical finding, since it reflects the left ventricular dysfunction. Causes of the disease are unknown, but alcoholism and viral infection are probably important precipitating factors.

- Heard best at cardiac apex.
- Mostly protodiastolic, especially with overt heart failure; presystolic also common, often accompanied by palpable "atrial kick"; may occur alone in absence of heart failure.
- Double or summation gallop sounds common in heart failure.
- When both gallop sounds occur, presystolic often louder than protodiastolic sound.

ASSOCIATED FINDINGS

- Progressive exertional dyspnea, orthopnea and paroxysmal nocturnal dyspnea.
- Nonproductive cough when heart failure present; vague chest pain at times.
- Cyanosis uncommon unless heart failure is severe.
- Blood pressure normal or narrow pulse pressure.
- Pulsus alternans common, even without heart failure.
- Apical impulse diffuse, often sustained.
- First sound soft; second sound normally split; P_2 loud in heart failure.
- Brief or holosystolic murmur at apex in heart failure; usually disappears with compensation.
- Bilateral moist rales reflect left heart failure; distended neck veins, hepatomegaly, edema reflect right heart failure.
- ECG may show left ventricular hypertrophy, nonspecific ST-T abnormalities or left bundle branch block.
- Atrial fibrillation common.
- X-ray will show general cardiomegaly.

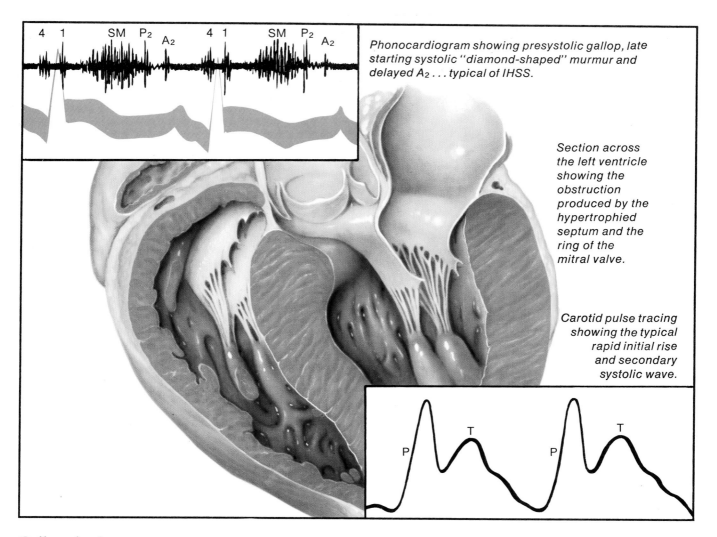

Phonocardiogram showing presystolic gallop, late starting systolic "diamond-shaped" murmur and delayed A_2 ... typical of IHSS.

Section across the left ventricle showing the obstruction produced by the hypertrophied septum and the ring of the mitral valve.

Carotid pulse tracing showing the typical rapid initial rise and secondary systolic wave.

Gallop rhythm in
IDIOPATHIC HYPERTROPHIC SUBAORTIC STENOSIS

Gallop rhythm may often be the only clearly abnormal auscultatory finding in idiopathic hypertrophic subaortic stenosis. This recently recognized disease is caused by a piling up of muscle below the aortic or pulmonic valve causing an obstruction to ventricular outflow. All physical and laboratory findings are variable; the condition usually simulates aortic stenosis. The disease should be suspected chiefly in children and young adults.

- Heard best at apex; sometimes also at xiphoid.
- Presystolic most frequent and most prominent, can be enhanced by infusion of isoproterenol; often accompanied by palpable "atrial kick" at apex.
- Apical protodiastolic gallop may be present, especially in heart failure; double sound or summation gallop at times.
- Prominent *a* wave in jugular pulse from associated right-sided subvalvular stenosis.

ASSOCIATED FINDINGS

- Exertional dyspnea common; orthopnea, paroxysmal nocturnal dyspnea uncommon.
- Exertional syncope frequent, chest pain at times.
- Blood pressure normal or narrow pulse pressure.
- Arterial pulse rises rapidly and collapses; secondary rise prior to dicrotic notch.
- Apical impulse forceful and sustained.
- First sound normal or loud; no ejection sounds; A_2 soft, often late with paradoxical splitting of second sound.
- Ejection murmur low along left sternal border; may have thrill; late systolic murmur at apex at times, with associated mitral insufficiency.
- Intensity of first sound, murmur and bifid character of pulse, as well as gallop, increased by isoproterenol infusion; intensity of ejection murmur decreased by vasopressors, increased by amyl nitrite inhalation.
- ECG shows left ventricular hypertrophy; at times Q waves suggesting myocardial infarction.
- X-ray shows cardiomegaly but no poststenotic dilatation of aorta.

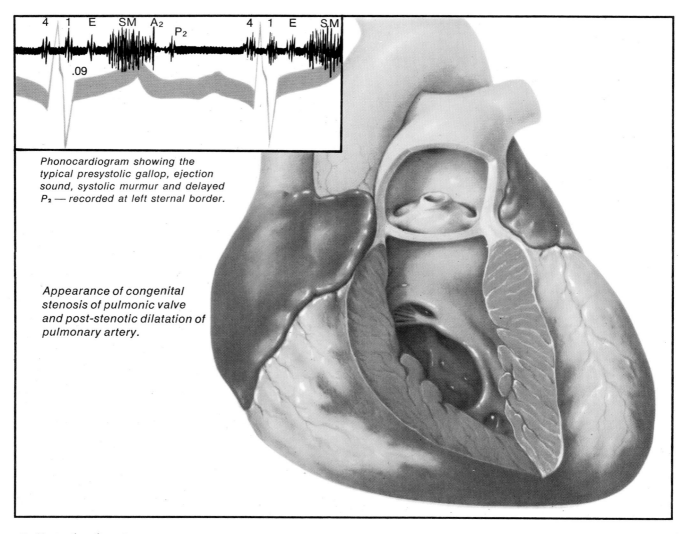

Phonocardiogram showing the typical presystolic gallop, ejection sound, systolic murmur and delayed P_2 — recorded at left sternal border.

Appearance of congenital stenosis of pulmonic valve and post-stenotic dilatation of pulmonary artery.

Gallop rhythm in PULMONIC STENOSIS

Gallop rhythm, although not as typical as systolic murmur in pulmonic stenosis, is of great value in assessing the severity of the condition and establishing the correct diagnosis. Ignorance of its occurrence in pulmonic stenosis may lead to a false diagnosis of a lesion of the left heart. The condition occurs rarely as a result of rheumatic fever or from a serotonin- and kallikrein-secreting argentaffinoma. Usually, however, it can be considered congenital in origin, though it may be discovered initially in adults as well as in children.

- At xiphoid and along lower left sternal border.
- Usually presystolic alone; when marked, may be associated with presystolic hepatic pulsation; may occur with or without heart failure.
- Often associated with prominent *a* waves in jugular venous pulse; gallop and large *a* wave not present with complicating ventricular septal defect
- Protodiastolic, double and summation gallops rare.

ASSOCIATED FINDINGS

- Exertional dyspnea or breathlessness.
- Cyanosis when atrial septum is patent.
- Hypertelorism and moon facies in extreme cases.
- Blood pressure normal or narrow pulse; post-Valsalva arterial pressure overshoot absent despite normal fall in pressure during straining.
- Left parasternal lift (right ventricular heave).
- Normal first sound, followed by ejection sound that increases on expiration; late and diminished pulmonic component of second sound; split second sound, may be fixed split in severe lesions.
- Ejection-type systolic murmur along left sternal border radiating to left shoulder.
- Distended neck veins, enlarged liver, edema with right heart failure.
- ECG shows right ventricular and right atrial hypertrophy, often with large late R in V_1.
- X-ray shows cardiomegaly and poststenotic dilatation of pulmonary artery, except with infundibular stenosis.

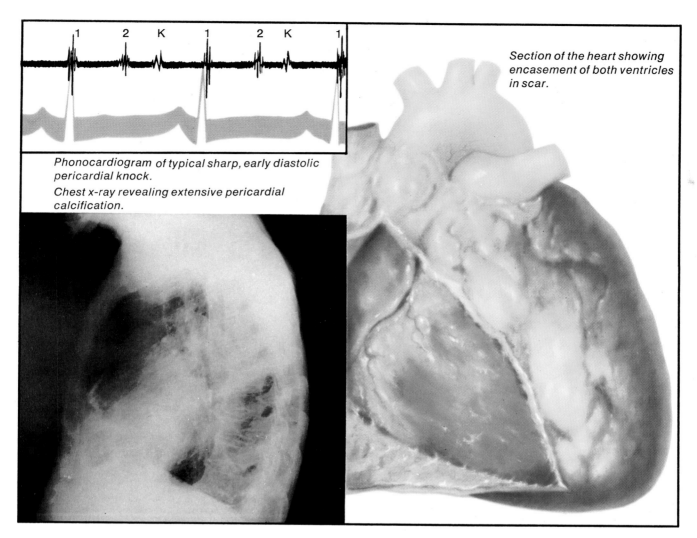

Phonocardiogram of typical sharp, early diastolic pericardial knock.

Chest x-ray revealing extensive pericardial calcification.

Section of the heart showing encasement of both ventricles in scar.

Gallop rhythm in
CONSTRICTIVE PERICARDITIS

A protodiastolic gallop rhythm of a special type (pericardial knock) is the most constant auscultatory finding and an important diagnostic clue in constrictive pericarditis. This chronic condition, which simulates heart failure, is produced by the gradual encasement of the heart in a rigid shell. Formerly, tuberculosis of the pericardium was the most common cause. Today, tuberculous pericarditis is uncommon and the cause is often unknown, but is probably usually viral.

- Can be in right or left ventricle; may be maximal at apex or left sternal border.
- Protodiastolic only; sharper and earlier than usual gallop (.06 to .12 second following aortic closure).
- Presystolic gallop may be present when constriction is limited to left ventricle and there is associated pulmonary hypertension.

ASSOCIATED FINDINGS

- Exertional dyspnea, weakness common; orthopnea, paroxysmal nocturnal dyspnea uncommon.
- Distended neck veins, hepatomegaly, ascites, edema occur early and persist.
- Sudden sharp collapse of jugular pulse — steep y descent.
- Normal blood pressure or narrowed pulse pressure; pulsus paradoxus at times; systemic arterial pressure falls more than 10 mm Hg with inspiration.
- Apical impulse often difficult to feel.
- First and second heart sounds usually distant.
- Cardiac murmur and pericardial friction rub rare; rales uncommon.
- Marked proteinuria, with nephrotic syndrome.
- ECG shows nonspecific ST-T abnormalities, right ventricular hypertrophy at times, low voltage QRS at times.
- Atrial fibrillation in one-third of cases; atrial flutter at times.
- X-ray shows small or slightly enlarged heart (due to pericardial thickening), pericardial calcification in 40 to 50%; definite diagnosis by cardiac catheterization and angiocardiogram.

DIFFERENTIAL DIAGNOSIS OF GALLOP RHYTHM

	Location	Presystolic	Protodiastolic	Double Sound
ARTERIOSCLEROTIC HEART DISEASE	Apex; "atrial kick" with aneurysm	In incipient or subsiding heart failure; with angina; with aneurysm	In heart failure — with atrial fibrillation or weak atrial contraction	In heart failure — with slower heart rate; summation gallop with faster rate
HYPERTENSIVE HEART DISEASE	Apex; with "atrial kick" at times	In incipient or subsiding heart failure; without CHF if arterial pressure is high	In heart failure — with atrial fibrillation	In heart failure — with slower heart rate; summation gallop with faster rate
AORTIC STENOSIS	Apex; often with "atrial kick"	In heart failure; or without it if valve obstruction is marked	At times occurs alone — in heart failure with atrial fibrillation	In heart failure — with slower heart rate; summation gallop with faster rate
AORTIC INSUFFICIENCY	Apex	Uncommon even with large regurgitation	Frequent with CHF; or without it if regurgitation is marked	Uncommon
MITRAL INSUFFICIENCY	Apical; at xiphoid with pulmonary hypertension	Rare in left heart; may occur with small atrium	Frequent with or without CHF; occurs earlier in diastole than other gallops	Infrequent even with CHF; if both gallop sounds occur, locations are different

Associated Signs and Symptoms	Key Laboratory Data
Dyspnea; orthopnea; paroxysmal nocturnal dyspnea; cough; chest pain; pulsus alternans; diminished first sound; paradoxical split of S_2; systolic murmur at apex; signs of left and right heart failure	ECG — evidence of infarction; x-ray may show cardiomegaly; fluoroscopy shows paradoxical motion with ventricular aneurysm, apical "atrial kick" with aneurysm
Dyspnea; orthopnea; paroxysmal nocturnal dyspnea; cough; headache; blurred vision; elevated arterial pressure; loud first sound and A_2; signs of left and right heart failure; arteriolar narrowing in retinal fundi	ECG — LVH, secondary ST-T abnormality; x-ray may show cardiomegaly, dilated aortic arch
Dyspnea, orthopnea, paroxysmal nocturnal dyspnea; cough; chest pain or syncope; narrow pulse pressure; small, slowly rising arterial pulse; pulsus alternans; systolic ejection murmur; reduced splitting of second sound; signs of left and right heart failure	ECG — LVH with secondary ST-T abnormality, LBBB or A-V block; x-ray may show cardiomegaly, dilatation of ascending aorta
Dyspnea, orthopnea, paroxysmal nocturnal dyspnea; awareness of heart beat, neck pulsation; sweating, heat intolerance; chest pain; prominent arterial pulses; first sound increased; decrescendo diastolic murmur at base; signs of left or right heart failure at times	ECG — left axis deviation; signs of LVH not consistent; x-ray shows cardiomegaly with elongated boot-shaped heart and dilated aorta
Dyspnea, orthopnea, paroxysmal nocturnal dyspnea; fatigue; cough; palpitation; small, rapidly rising and falling pulse; first sound often obscured by holosystolic apical murmur; signs of left and right heart failure at times; *a* wave prominent with pulmonary hypertension	ECG — LVH, P mitrale when rhythm is sinus, atrial fibrillation at times; x-ray shows enlargement of left ventricle and atrium, at times of pulmonary arteries

(Continued on next page.)

DIFFERENTIAL DIAGNOSIS OF GALLOP RHYTHM (continued)

	Location	Presystolic	Protodiastolic	Double Sound
PULMONARY HYPERTENSION	At xiphoid and/or along lower sternal border	With or without heart failure when pulmonary hypertension is marked	Uncommon even with marked CHF	Rare
PRIMARY MYOCARDIAL DISEASE	Apex; often with "atrial kick"	Frequent; often alone in absence of CHF	Most common, especially when heart failure is present	Common in CHF with slower rate; summation gallop with faster rate
IDIOPATHIC HYPERTROPHIC SUBAORTIC STENOSIS	Apex, may be at xiphoid, often with "atrial kick"	Frequent with or without CHF	May be present, especially with heart failure; usually not as loud as presystolic	Occurs in CHF with slower rate; summation gallop with faster rate
PULMONIC STENOSIS	At xiphoid; may be at lower left sternal border	Frequent with or without CHF	Rare	Rare
CONSTRICTIVE PERICARDITIS	Apical; at times also at xiphoid	Rare	Most common; sharper and earlier than usual gallop — pericardial knock	Rare

Associated Signs and Symptoms	Key Laboratory Data
Exertional or resting dyspnea from pulmonary disease; cyanosis; cough; BP and pulse normal or diminished; right ventricular heave; loud P_2; ejection sound at times; signs of right ventricular failure; jugular *a* wave prominent unless atrial fibrillation is present	ECG — RVH, often atrial fibrillation; x-ray will show large pulmonary arteries and characteristic findings of primary disease
Progressive exertional dyspnea, orthopnea, paroxysmal nocturnal dyspnea; cough; vague chest pains; BP normal or narrow pulse pressure; pulsus alternans; first sound soft; P_2 loud in CHF with brief or holosystolic murmur at apex; signs of left and right heart failure	ECG may show LVH or LBBB, or nonspecific ST-T abnormalities; x-ray will show cardiomegaly; apical "atrial kick" frequent
Exertional dyspnea; exertional syncope; chest pain at times; blood pressure normal or narrow pulse pressure; rapid rise, then fall and second rise of arterial pulse; first sound loud; ejection murmur along lower left sternal border; prominent *a* wave in neck veins frequent	ECG shows LVH, and at times abnormal Q waves; x-ray will show cardiomegaly but no post-stenotic aortic dilatation; gallop increased by isoproterenol
Exertional dyspnea; breathlessness; cyanosis; left parasternal lift; first sound normal; ejection sound; ejection systolic murmur along left sternal border; soft P_2; signs of right heart failure; prominent jugular *a* wave frequent	ECG — RVH and right atrial hypertrophy with large, late R in V_1; x-ray shows cardiomegaly and post-stenotic dilatation of pulmonary artery
Exertional dyspnea; weakness; hepatomegaly; ascites; edema; distended neck veins; normal BP or narrow pulse pressure; pulsus paradoxus; heart sounds soft; rales uncommon	ECG — nonspecific ST-T abnormalities; RVH at times; atrial fibrillation, flutter at times; x-ray will show normal or slightly enlarged heart, often with calcification

SYSTOLIC MURMUR

Murmurs in systole — additional sounds of some duration occurring between the first and second heart sounds — are so common in a variety of diseases and even in the normal subject that the simple identification of such a murmur is of little diagnostic value per se.

On the other hand, when the murmur is accurately characterized as to its location, radiation, intensity, timing, duration and quality; when its response to respiration, position and to certain drugs is assessed; and when it is correlated with other cardiac signs and symptoms, it then becomes a valuable tool in differential diagnosis.

The systolic murmurs characteristic of twelve important conditions and the principal signs and symptoms of these conditions are described and then compared in tabular form.

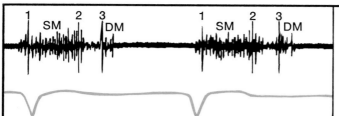

Typical holosystolic murmur of mitral insufficiency. In the case shown a loud first sound is recorded.

An angiocardiogram recorded in the left anterior oblique projection showing radiopaque dye injected into the left ventricle regurgitating into the left atrium.

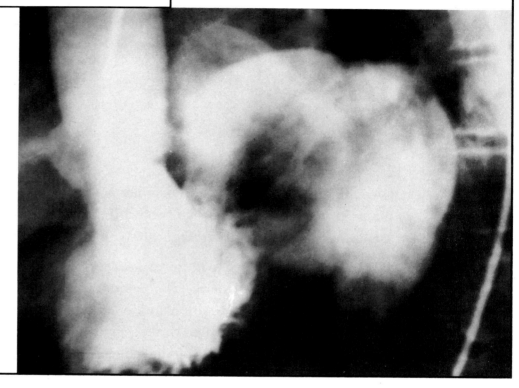

Angiocardiogram recorded in the left anterior oblique projection showing radio-opaque dye injected into the left ventricle regurgitating into the left atrium.

Systolic Murmur due to
MITRAL INSUFFICIENCY

The systolic murmur of mitral insufficiency is usually quite loud; however, variations in this murmur and similarities to the murmurs that can occur in the normal heart or in heart failure due to many causes make the analysis of the associated findings very important in establishing the diagnosis.

- Maximal at cardiac apex or just lateral to it.
- Often radiates widely, especially to left axilla.
- Thrill common at cardiac apex and in left axilla.
- Typically holosystolic . . . begins with first sound and lasts to or slightly past A_2 and even P_2.
- Intensity of murmur can be increased by phenylephrine, reduced by amyl nitrite.
- Murmur harsh, nonmusical and constant in intensity.
- Occasionally, the murmur may be present only in early systole, or only in late systole following a mid-systolic click.
- Mid-diastolic rumble without presystolic accentuation often present, need not imply mitral stenosis.

ASSOCIATED FINDINGS

- Dyspnea, orthopnea, paroxysmal nocturnal dyspnea with heart failure; may be preceded by fatigue and palpitation.
- Little water hammer pulse, with rapid rise and fall.
- Vigorous apical impulse.
- First sound normal or loud, but often obscured by murmur; diminished with advanced failure.
- Protodiastolic gallop, often close to second sound, with or without heart failure.
- Presystolic gallop usually only from right heart with pulmonary hypertension.
- Opening snap at times — with mobile anterior leaflet.
- Atrial fibrillation frequent.
- ECG — often LVH, occasionally only RVH; often nondescript; P mitrale when rhythm is sinus.
- X-ray — enlarged left ventricle, left atrium; at times pulmonary artery and right ventricle also enlarged.

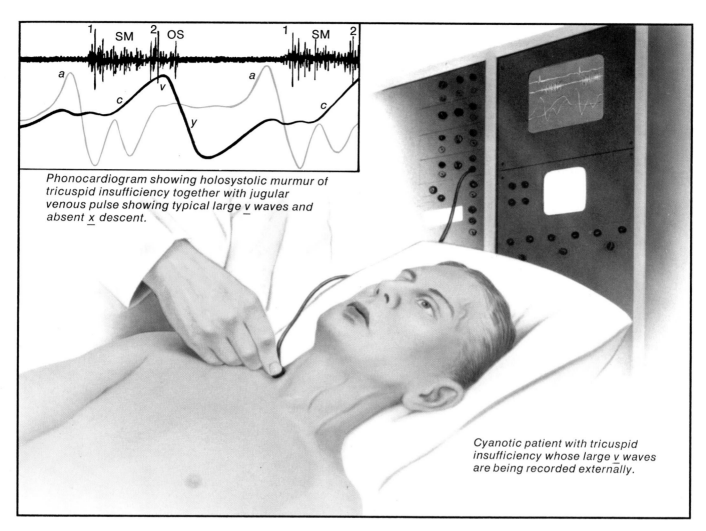

Phonocardiogram showing holosystolic murmur of tricuspid insufficiency together with jugular venous pulse showing typical large v waves and absent x descent.

Cyanotic patient with tricuspid insufficiency whose large v waves are being recorded externally.

Systolic murmur due to TRICUSPID INSUFFICIENCY

Insufficiency of the tricuspid valve due to organic disease of the valve — whether rheumatic, congenital or serotonin-induced — is quite uncommon. Most often the murmur of tricuspid insufficiency is due to functional incompetence secondary to heart failure, and is important primarily because it may cause confusion with other lesions.

- Maximal at lower end of sternum or in 4th left intercostal space; radiates to right sternal border, also to apex with marked right ventricular dilatation.
- Thrill may be present at lower end of sternum.
- Holosystolic; onset obscures first sound; murmur continues to or through second sound.
- Harsh, nonmusical and constant in intensity.
- Murmur often increased by deep inspiration; otherwise, associated tricuspid stenosis must be suspected.
- Short mid-diastolic rumble with associated tricuspid stenosis — functional or organic.

ASSOCIATED FINDINGS

- In rare pure tricuspid insufficiency symptoms only of right heart failure — abdominal swelling and edema.
- Dyspnea on exertion common from associated mitral or pulmonary disease.
- Orthopnea and paroxysmal nocturnal dyspnea less than expected, with left heart disease.
- Cyanosis and/or icterus occasionally; hepatomegaly, ascites, edema frequently.
- Marked venous engorgement.
- Jugular venous pulse prominent with big v and y descent; absence of x descent.
- See-saw motion of chest wall, systolic hepatic expansion together with cardiac depression.
- Protodiastolic gallop from right side usually not present.
- Atrial fibrillation common.
- Prominent P waves on ECG when atrial fibrillation absent; ECG evidence of RVH often present.
- Characteristic x-ray finding is right atrial enlargement; other findings due to underlying disease.

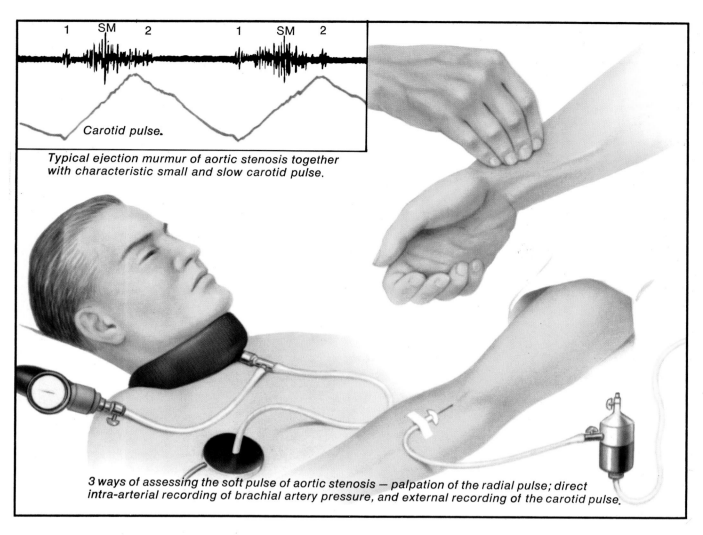

Typical ejection murmur of aortic stenosis together with characteristic small and slow carotid pulse.

3 ways of assessing the soft pulse of aortic stenosis — palpation of the radial pulse; direct intra-arterial recording of brachial artery pressure, and external recording of the carotid pulse.

Systolic murmur due to
VALVULAR AORTIC STENOSIS

The harsh systolic murmur of aortic stenosis is the most characteristic feature of this condition, and is easily heard. The difficulty in diagnosis is the differentiation of this murmur from those due to other cardiac lesions, and from murmurs of no dynamic importance, which are frequent in the same area, especially in older patients.

- Maximal at base of heart, usually second intercostal space to the right of sternum.
- Radiates to neck, upper back, apex and often to right axilla.
- Thrill frequent in second right interspace, suprasternal notch and over carotids.
- Begins just after first sound, ends before aortic component of second; the more severe the obstruction, the later the onset of murmur.
- Harsh ejection (crescendo-decrescendo) murmur.
- A faint murmur of aortic insufficiency frequent.

ASSOCIATED FINDINGS

- Dyspnea, orthopnea, paroxysmal nocturnal dyspnea, nonproductive cough with heart failure.
- Pressing substernal pain or syncope on exertion.
- Arterial pulses soft and slowly rising.
- Pulsus alternans often present.
- Forceful and sustained apical impulse.
- Ejection sound frequent in milder cases.
- Apical presystolic gallop frequent, often with palpable "atrial kick"; protodiastolic or double gallop with heart failure.
- First sound normal or loud; A_2 often obscured and reduced in intensity; splitting of second sound reduced, paradoxical splitting may occur.
- Moist rales with left heart failure; distended veins, hepatomegaly and edema with right heart failure.
- ECG — left ventricular hypertrophy with secondary ST-T abnormality of the "systolic overload" type; may show bundle branch block or complete block when calcification extends into septum.
- X-ray shows cardiomegaly and dilatation of ascending aorta; aortic valve calcification seen in most adults, when image intensification used.

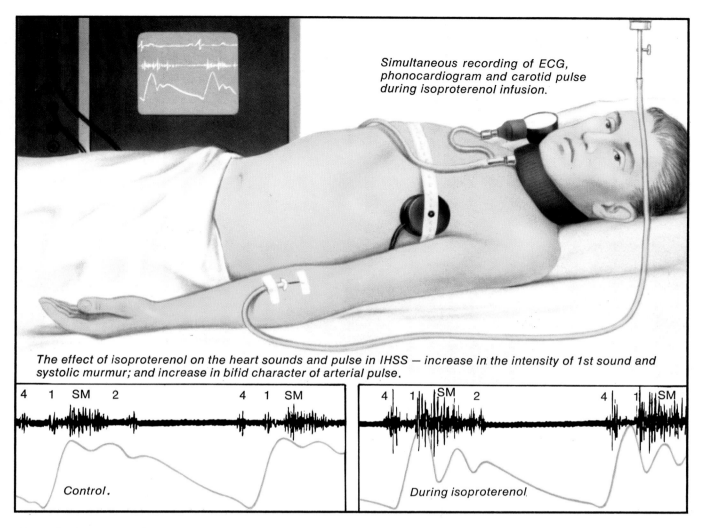

Simultaneous recording of ECG, phonocardiogram and carotid pulse during isoproterenol infusion.

The effect of isoproterenol on the heart sounds and pulse in IHSS — increase in the intensity of 1st sound and systolic murmur; and increase in bifid character of arterial pulse.

Control.

During isoproterenol.

Systolic Murmur due to
IDIOPATHIC HYPERTROPHIC SUBAORTIC STENOSIS

The systolic murmur of IHSS, like the muscular obstruction it reflects, can be quite variable. In some instances it mimics exactly the murmur of aortic stenosis. At other times it is quite nondescript. Any basal murmur, especially in a child or younger adult, requires a search for other findings present in this disease.

- Maximal at left sternal border, usually in 3rd and 4th intercostal space.
- Radiates to apex, not usually to neck.
- Thrill at times, maximal at lower left sternal border.
- Tends to begin later after 1st sound than ordinary aortic stenosis; ends before A_2.
- Sometimes ejection (crescendo-decrescendo) in character, sometimes nondescript.
- Often pansystolic because of superimposed mitral insufficiency, especially at apex.
- Diastolic blowing murmur of aortic insufficiency absent.
- Murmur may be decreased by methoxamine or phenylephrine, but may be increased by amyl nitrite inhalation.

ASSOCIATED FINDINGS

- Exertional dyspnea common; orthopnea, paroxysmal nocturnal dyspnea uncommon.
- Exertional syncope frequent; chest pain at times.
- Arterial pulse rises rapidly and collapses with secondary rise prior to dicrotic notch.
- First sound normal or loud; no ejection sound.
- A_2 soft, often late, with paradoxical splitting of second sound.
- Presystolic gallop at apex with "atrial kick" frequent; protodiastolic gallop at times, especially in heart failure.
- May have prominent *a* wave in neck from associated right-sided lesion.
- Gallop, intensity of first sound, murmur and bifid character of pulse all increased by isoproterenol infusion.
- ECG shows LVH, at times Q waves suggesting myocardial infarction.
- X-ray will show cardiomegaly but no poststenotic dilatation of aorta.

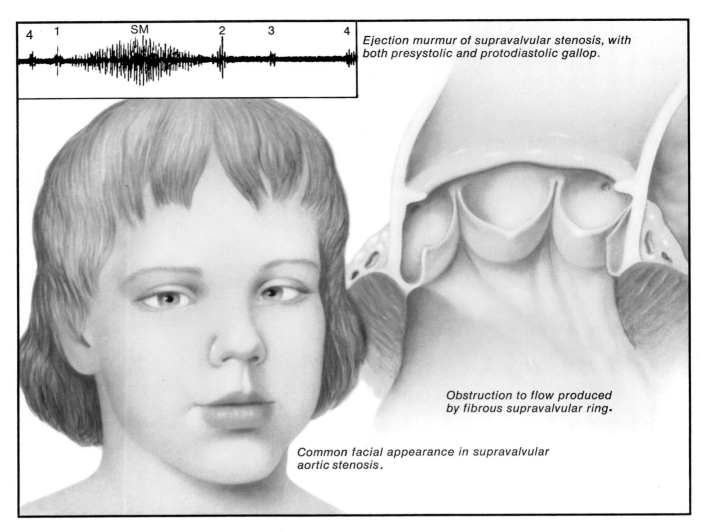

Ejection murmur of supravalvular stenosis, with both presystolic and protodiastolic gallop.

Obstruction to flow produced by fibrous supravalvular ring.

Common facial appearance in supravalvular aortic stenosis.

Systolic murmur due to
SUPRAVALVULAR AORTIC STENOSIS

The murmur of supravalvular aortic stenosis — except, at times, for a difference in location — is quite like that of aortic stenosis. This congenital, often familial, defect, however, is unique in that there are often other characteristic noncardiac features: typical facies and often a history of idiopathic hypercalcemia in infancy.

- Often maximal over suprasternal notch and right side of neck, sometimes in second right intercostal space.
- Radiates to neck, less so to apex.
- Thrill at times, often in suprasternal notch.
- Begins after first sound, ends before second.
- Harsh ejection type (crescendo-decrescendo).
- Soft diastolic blowing murmur of aortic insufficiency in about one-fourth of patients.

ASSOCIATED FINDINGS

- Familial in one-third of cases.
- Exertional dyspnea, syncope and chest pain, as in other forms of aortic stenosis.
- Characteristic facial appearance at times—fullness of lower lip and cheeks, prominence of nose and chin.
- Strabismus common.
- Mental and physical retardation in some cases.
- Slow and small arterial pulse.
- Blood pressure in right arm often 20 mm Hg, or more, higher than in left.
- First sound normal or loud.
- Ejection sound does not occur.
- Presystolic gallop may be present.
- Signs and symptoms of left and right heart failure may occur.
- ECG shows LVH and often very marked ST and T abnormalities — probably from ischemia.
- X-ray shows moderate cardiomegaly, no poststenotic dilatation of aortic root and a small aortic knob.
- Multiple pulmonary coarctations may be associated.
- Diagnosis confirmed by demonstration of supravalvular chamber by angiography and left heart catheterization.

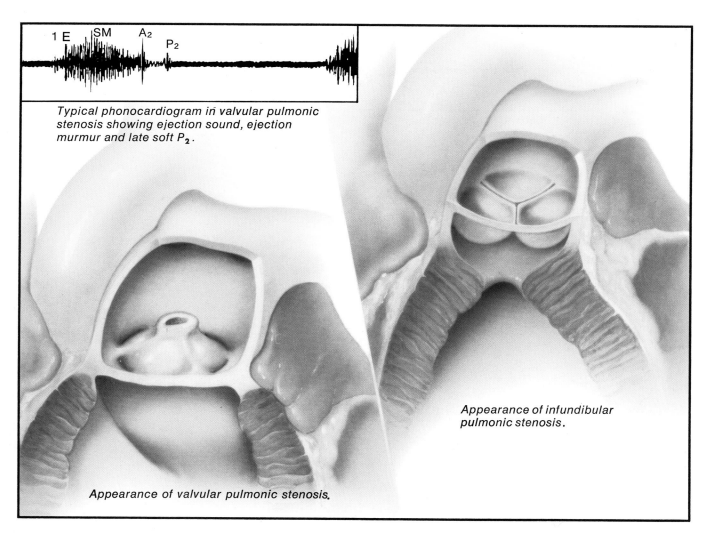

Typical phonocardiogram in valvular pulmonic stenosis showing ejection sound, ejection murmur and late soft P_2.

Appearance of infundibular pulmonic stenosis.

Appearance of valvular pulmonic stenosis.

Systolic murmur due to
PULMONIC STENOSIS

A systolic murmur is usually present from infancy in pulmonic stenosis, since — whether valvular, infundibular or supravalvular — the condition is usually congenital, though it may not be detected until adult life. The majority are valvular; about 10% are infundibular; and a very few are supravalvular, often associated with multiple pulmonary coarctations.

- Murmur of valvular stenosis maximal in 2nd LICS, sometimes an interspace lower in infundibular.
- Radiates to left side of neck and sometimes left shoulder posteriorly; often radiates diffusely over lungs with supravalvular stenosis.
- Thrill common, usually in 2nd LICS, close to sternum.
- With more obstruction, typical harsh ejection (crescendo-decrescendo) murmur — the greater the obstruction, the later the peak of murmur; nondescript with little obstruction.
- Begins after first sound; may extend almost to P_2.

ASSOCIATED FINDINGS

- Exertional dyspnea or breathlessness; no orthopnea or paroxysmal nocturnal dyspnea.
- Cyanosis when atrial septum is patent.
- Blood pressure normal or pulse narrow.
- Left parasternal lift — right ventricular heave.
- First sound normal; second sound split, may be fixed split with severe lesion; P_2 late and soft with valvular stenosis, normal intensity with infundibular stenosis, sometimes loud with supravalvular stenosis.
- Ejection sound usually present; increases with expiration in mild cases, may not vary in moderate, absent in severe cases.
- Distended neck veins, enlarged liver, edema with right heart failure.
- Atrial fibrillation common; ventricular rate often rapid and difficult to control.
- ECG shows right ventricular and right atrial hypertrophy, often with large late R in V_1.
- X-ray: cardiomegaly and poststenotic dilatation of pulmonary artery, except with infundibular stenosis.

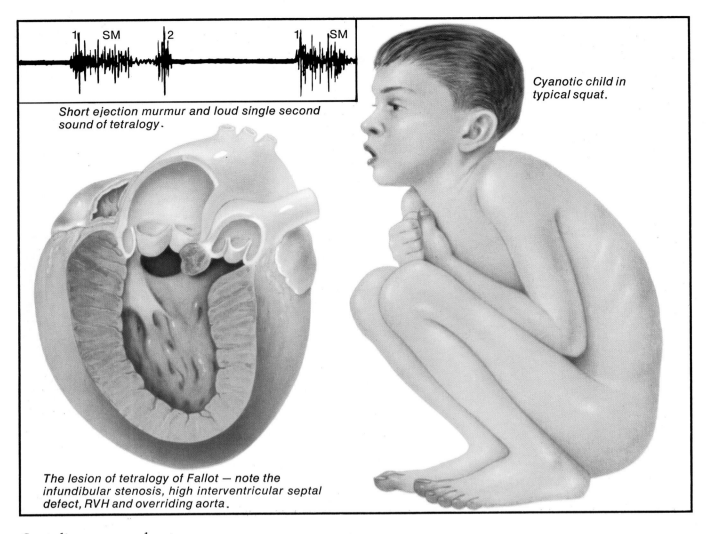

Short ejection murmur and loud single second sound of tetralogy.

Cyanotic child in typical squat.

The lesion of tetralogy of Fallot — note the infundibular stenosis, high interventricular septal defect, RVH and overriding aorta.

Systolic murmur due to TETRALOGY OF FALLOT

The murmur of tetralogy of Fallot does not usually present problems of differential diagnosis with the other entities compared here, since it generally occurs in cyanotic infants or children. At times, however, patients with this congenital condition can survive into adult life, and may be cyanotic only on exercise. These are patients with combined pulmonic stenosis and ventricular septal defect rather than patients with classical tetralogy.

- Maximal in 2nd or 3rd left ICS at LSB.
- May or may not radiate to cervical vessels.
- Thrill in 3rd left intercostal space at left sternal border in 50%.
- Begins after 1st sound, ending well before 2nd sound.
- Harsh ejection type (crescendo-decrescendo) murmur, peaks prior to midsystole.
- With more severe stenosis, murmur is softer; continuous murmur of patent ductus or bronchial collaterals may be heard.
- Murmur shorter than in simple pulmonic stenosis, earlier peak, reduced in intensity by amyl nitrite.

ASSOCIATED FINDINGS

- Cyanosis almost always present after 6 mos.; intensified with exercise, infection, advancing disease.
- Dyspnea, easy fatigability and retarded growth and development present early.
- Squatting common between 1½ and 8 years.
- Hypoxic spells with increased dyspnea and syncope in severe cases.
- Clubbing prominent.
- Loud single second heart sound present.
- Presystolic gallop and prominent jugular *a* wave of simple pulmonic stenosis are absent.
- Signs of congestive failure not present without complicating lesion.
- ECG shows right axis deviation and tall R in V_1, reflecting right ventricular hypertrophy.
- X-ray shows normal-sized heart on PA view, but right ventricular enlargement on lateral view; pulmonary conus is small and pulmonary vascularity decreased.

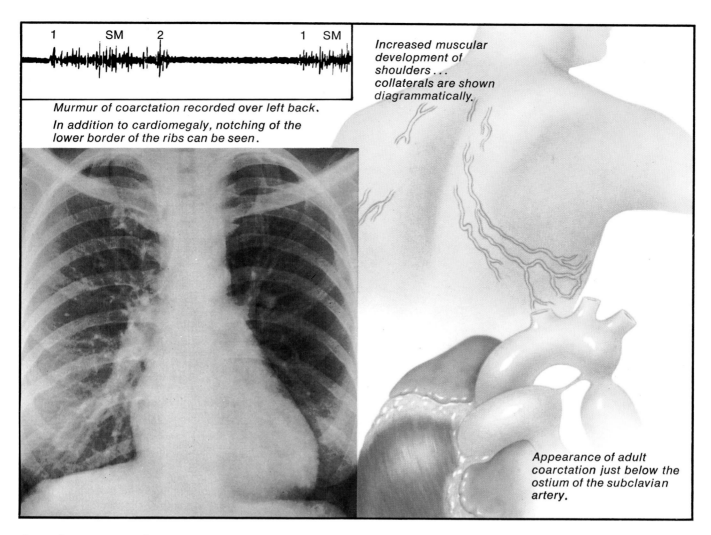

Murmur of coarctation recorded over left back.

In addition to cardiomegaly, notching of the lower border of the ribs can be seen.

Increased muscular development of shoulders... collaterals are shown diagrammatically.

Appearance of adult coarctation just below the ostium of the subclavian artery.

Systolic murmur due to
COARCTATION OF THE AORTA

A systolic murmur is usually present in coarctation of the aorta and may be the first clue that hypertension in an asymptomatic patient is due to this congenital defect. Although twice as common in men as in women, it occurs frequently in ovarian agenesis (Turner's syndrome).

- Maximal over left mid-back between scapula and spine.
- Radiates to neck and upper precordium, especially at base.
- Thrill may be present in back and in suprasternal notch; not usually in precordium.
- Begins after first sound, continues to or beyond second sound in more severe cases.
- Ejection murmur of aortic stenosis in 10%, diastolic basal murmur of aortic insufficiency in 30% — both due to associated bicuspid aortic valve.
- Systolic bruits, at times with thrill, often present over intercostal collaterals.

ASSOCIATED FINDINGS

- Often asymptomatic, may be discovered because of hypertension or murmur.
- At times claudication or weakness in legs.
- First symptoms may be those of bacterial endocarditis, dissecting aneurysm or cerebral hemorrhage.
- Symptoms of congestive failure, either in first year or after third decade.
- Persistent headaches may occur.
- Excessive development of arm and shoulder muscles; legs often relatively thin and underdeveloped.
- Blood pressure elevated in arms; may be 20 to 30 mm Hg, or more, higher than in legs; pulse in legs soft and late compared to arms; pulse in left wrist soft at times when left subclavian is involved in coarctation.
- A_2 often accentuated.
- Collateral vessels often felt between ribs.
- ECG shows high voltage QRS in half of cases; typical ST-T changes of LVH uncommon.
- X-ray shows some cardiomegaly and, after age 6, notching under edge of ribs; indentation of coarctation itself can be seen at times.

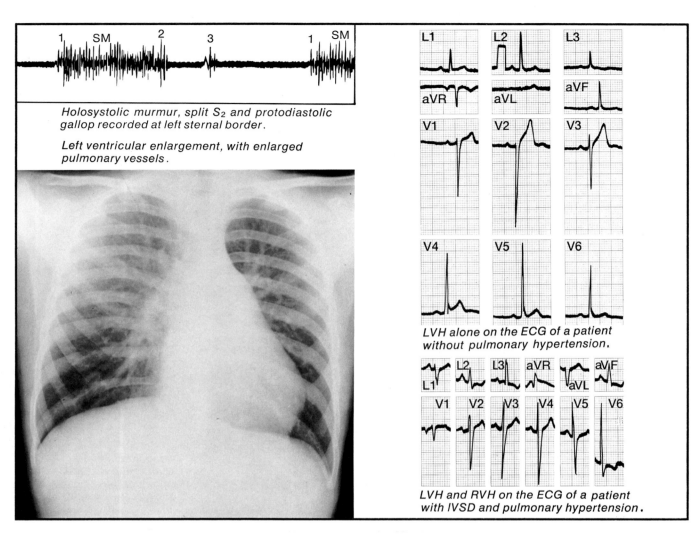

Holosystolic murmur, split S_2 and protodiastolic gallop recorded at left sternal border.

Left ventricular enlargement, with enlarged pulmonary vessels.

LVH alone on the ECG of a patient without pulmonary hypertension.

LVH and RVH on the ECG of a patient with IVSD and pulmonary hypertension.

Systolic Murmur due to
INTERVENTRICULAR SEPTAL DEFECT

The murmur produced by a defect in the interventricular septum is usually quite loud, even when the defect is slight and lacking in hemodynamic significance. With such defects the diagnosis may have to be made on the basis of the typical murmur alone. With larger defects, the associated findings will clarify the diagnosis. While usually congenital, IVSD must be considered a possibility in patients after myocardial infarction, when it can result from rupture of the septum.

- Maximal along lower sternal border.
- Radiates widely — often to lower thorax posteriorly . . . not usually to cervical vessels.
- Thrill frequent — in about ⅔ of cases — along lower left sternal border.
- Holosystolic murmur, beginning after 1st sound, may extend slightly past A_2; usually constant in intensity; occasionally crescendo-decrescendo.
- With a large shunt, there may also be an early diastolic or mid-diastolic rumble at apex.
- With severe pulmonary hypertension, the basal diastolic blowing murmur of pulmonic insufficiency may occur.

ASSOCIATED FINDINGS

- Small defects may be asymptomatic.
- Mild exertional dyspnea is early symptom.
- Large shunt can cause impaired growth, dyspnea, heart failure; if pulmonary hypertension is marked, cyanosis may occur.
- Increased splitting of second sound; but split varies normally with respiration.
- With large shunt or moderately increased pulmonary pressure — increased splitting and loud P_2.
- With marked pulmonary hypertension and reduced shunt — split is lost and P_2 very loud.
- With moderate and large shunt, protodiastolic gallop will be heard.
- With marked pulmonary hypertension, ejection click and ejection murmur in pulmonic region.
- ECG: normal with small defect; will show LVH with large shunt; biventricular hypertrophy when significant pulmonary hypertension is present.
- X-ray may be normal or show enlarged left ventricle, left atrium and pulmonary arteries; with pulmonary hypertension, right ventricle may also be enlarged.

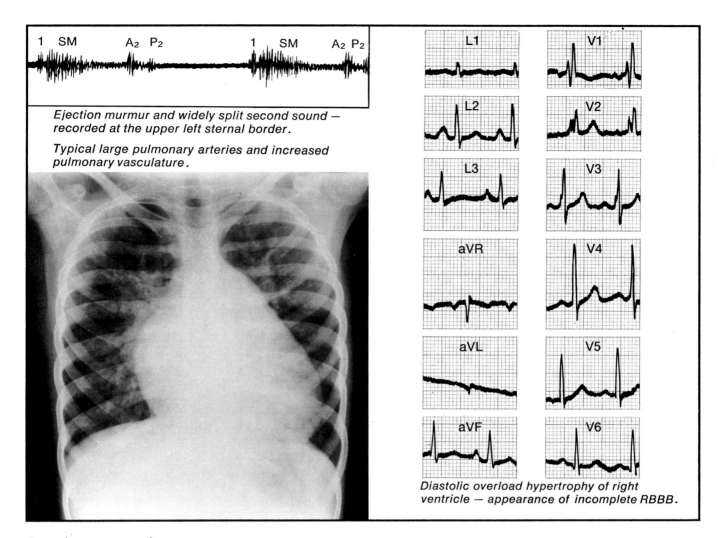

Ejection murmur and widely split second sound — recorded at the upper left sternal border.

Typical large pulmonary arteries and increased pulmonary vasculature.

Diastolic overload hypertrophy of right ventricle — appearance of incomplete RBBB.

Systolic murmur due to
INTERATRIAL SEPTAL DEFECT

The systolic murmur of interatrial septal defect is not unique or characteristic, since it is produced by turbulent flow in the pulmonary artery and is similar to a murmur common in normal children and in high output states. It may, however, warn of the possibility of heart disease in an asymptomatic patient and, together with associated findings, lead to the correct diagnosis.

- Maximal in second or third intercostal space along left sternal border.
- Usually localized; may radiate to apex.
- Begins after first sound, ends well before second.
- Harsh crescendo-decrescendo, peaks prior to midsystole.
- Diastolic rumble in about 1/3 of cases, from increased tricuspid flow, may be heard at apex or left sternal border; apical holosystolic murmur at times, from associated cleft mitral valve.

ASSOCIATED FINDINGS

- Exertional dyspnea and frequent respiratory infections with larger shunts.
- Right heart failure and cyanosis reflect advanced disease; rarely occur before third decade.
- Slender body build common.
- Left parasternal thrust.
- Second sound widely split, does not vary with respiration, reduced with pulmonary hypertension; P_2 may be loud and palpable even without pulmonary hypertension.
- Ejection sound in pulmonary artery at times; presystolic gallop at apex not uncommon.
- ECG shows right ventricular hypertrophy of diastolic overload type, indistinguishable from incomplete RBBB; when ECG is complicated by left axis deviation, ostium primum defect is likely; P-R interval prolonged in about 20% of cases; atrial fibrillation a late occurrence, as is complete right bundle branch block.
- X-ray shows cardiomegaly, enlarged pulmonary arteries and relatively small aortic knob.
- Cardiac fluoroscopy shows increased vascular pulsations — "hilar dance."

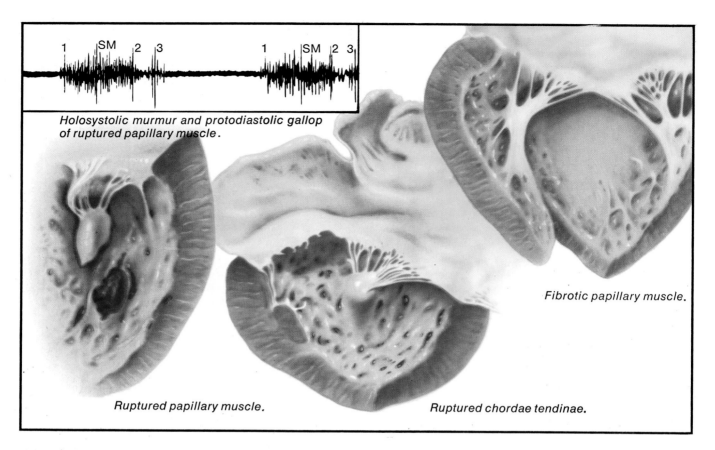

Holosystolic murmur and protodiastolic gallop of ruptured papillary muscle.

Ruptured papillary muscle.

Fibrotic papillary muscle.

Ruptured chordae tendinae.

Systolic murmur due to
DISEASE OF PAPILLARY MUSCLES OR CHORDAE TENDINAE

Systolic murmurs due to damage to the supporting structure of the mitral valve can be produced in several different ways, each form having characteristic features. Such murmurs can result from rupture of the papillary muscle (a rare and serious complication of acute myocardial infarction or trauma), from dysfunction and scarring of a papillary muscle as a result of infarction, or from rupture of the chordae tendinae usually following bacterial endocarditis but at times occurring spontaneously.

- Onset usually sudden with rupture; may be insidious with muscle dysfunction or with progressive rupture of chordae tendinae.
- Loud harsh murmur begins with S_1, to or just beyond S_2 — with ruptured papillary muscle.
- Usually ejection type (crescendo-decrescendo) or constant in intensity, beginning later in systole — with papillary muscle dysfunction.
- Holosystolic, late systolic or with late systolic accentuation — with ruptured chordae tendinae.
- Maximal at apex in all; radiates to axilla, sometimes to sternum with rupture.
- Usually no palpable thrill.

ASSOCIATED FINDINGS

Ruptured papillary muscle
- Severe dyspnea, signs of pulmonary edema, often shock.
- Protodiastolic gallop common.
- If patient survives acute onset, severe heart failure develops, often with holosystolic murmur of tricuspid insufficiency at left sternal border.
- Most likely with inferior or anterolateral infarcts — unlike ruptured septum, which occurs more often with anteroseptal infarcts.

Papillary muscle dysfunction
- More common but less acute than ruptured papillary muscle; regurgitation quantitatively less.
- May worsen failure following infarction.

Ruptured chordae tendinae
- Signs of pulmonary edema and heart failure may appear or be exacerbated when murmur appears.
- Loud, sharp apical protodiastolic gallop common; in some presystolic gallop at times.
- Often evidence of bacterial endocarditis.

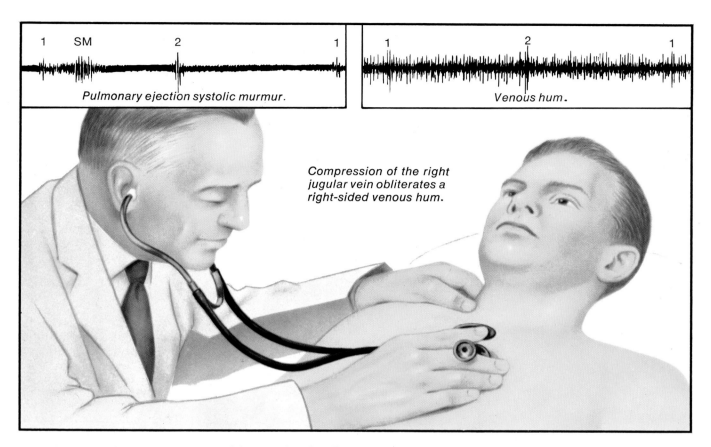

MURMURS IN THE NORMAL SUBJECT

Murmurs in the normal subject, or innocent murmurs, are varied in character but usually fall into one of several patterns which allow identification. Some occur only in systole; others are continuous throughout systole and diastole. Similar murmurs may occur in subjects who are not truly normal but in whom the murmurs result from increased flow rather than from an abnormality of the heart. In any case, all cardiac findings except the murmur are normal.

Venous hum — Best heard just above clavicle; often radiates to 2nd and 3rd interspaces.

- Continuous, usually louder in diastole.
- Loudest when sitting and when head is turned away from side of murmur; decreased by lying down; obliterated by compression of cervical veins.
- Can occur at any age, mostly between 5 and 10.
- Often mistaken for patent ductus arteriosus.

Mammary souffle — Common in 2nd intercostal space.

- Continuous, with diastolic accentuation.
- Can be obliterated by local pressure.
- Can be heard in last trimester of pregnancy and for several weeks postpartum.

Supraclavicular bruit — Maximal in supraclavicular space; usually on right, at times bilateral; may be transmitted to 1st and 2nd interspaces.

- Crescendo-decrescendo, peak may be early or late.
- Mostly in children or young adults.
- Sometimes mistaken for aortic or pulmonic stenosis.

Musical apical systolic murmur — Maximal at apex; radiates to left sternal border or base of heart; occasional thrill at apex.

- Usually holosystolic; markedly affected by respiration and posture; musical quality.
- Common in children.

Pulmonary ejection systolic murmur — Maximal in 2nd left intercostal space; no thrill.

- Begins after first, ends before second sound; crescendo-decrescendo, peak before mid-systole.
- Decreased by sitting up.
- Common in children or when chest wall is thin; basically same murmur in increased flow states — exercise, hyperthyroidism, fever, anemia, etc.

DIFFERENTIAL DIAGNOSIS OF SYSTOLIC MURMUR

	Timing	Location	Radiation	Quality	Thrill
MITRAL INSUFFICIENCY	Holosystolic; from 1st sound to 2nd or just beyond	At apex, or just lateral	Wide, especially to left axilla	Harsh, nonmusical, of constant intensity	Common at apex and/or left axilla
TRICUSPID INSUFFICIENCY	Holosystolic; from 1st sound to 2nd or just beyond	At lower end of sternum or 4th left intercostal space	To right sternal border ... also to apex	Harsh, nonmusical, of constant intensity; increased by deep inspiration	May be present at lower end of sternum
VALVULAR AORTIC STENOSIS	Begins just after 1st sound, ends before aortic component of 2nd sound	Usually 2nd intercostal space to right of sternum	To neck, upper back, apex and often right axilla	Harsh ejection (crescendo-decrescendo); greater obstruction with later peak	Frequent; in 2nd right interspace, in suprasternal notch and over carotids
IDIOPATHIC HYPERTROPHIC SUBAORTIC STENOSIS	Begins after 1st sound, ends before A_2	Maximal at lower left sternal border, usually in 3rd and 4th intercostal spaces	To apex, not usually to neck	Harsh ejection (crescendo-decrescendo) at times, but occasionally nondescript	Occasionally present; maximal at lower left sternal border
SUPRAVALVULAR AORTIC STENOSIS	Begins after 1st sound, ends before 2nd sound	Often over suprasternal notch and right side of neck; sometimes in right intercostal space	Well to neck, less so to apex	Harsh ejection (crescendo-decrescendo); greater obstruction — later peak	Often present in suprasternal notch
PULMONIC STENOSIS	Begins after 1st sound, may extend almost to P_2	Valvular usually in 2nd left intercostal space; infundibular often an interspace lower	To left side of neck, at times to left shoulder posteriorly	Harsh ejection (crescendo-decrescendo); greater obstruction — later peak	Common, usually in 2nd intercostal space, close to sternum
TETRALOGY OF FALLOT	Begins after 1st sound, ends well before 2nd sound	In 2nd or 3rd left intercostal space close to sternum	To cervical vessels, at times	Harsh ejection (crescendo-decrescendo); peak relatively early	Present in 50% of cases; in 3rd intercostal space at left sternal border
COARCTATION OF THE AORTA	Begins after 1st sound; continues to or beyond 2nd sound	Left mid-back between scapula and spine	To neck and precordium, especially at base	Harsh ejection (crescendo-decrescendo)	In back and in suprasternal notch; usually not in precordium

Associated Auscultatory Findings	Associated Signs and Symptoms	Key Laboratory Data
Mid-diastolic rumble without presystolic accentuation often present; 1st sound often obscured; protodiastolic gallop frequent	Fatigue, palpitation, symptoms of left heart failure; brisk small pulse	ECG — often LVH, or at times RVH; P mitrale or atrial fibrillation; x-ray — enlarged left ventricle and atrium, plus dilatation of pulmonary artery and right ventricle at times
May show short mid-diastolic rumble; usually no protodiastolic gallop	Symptoms of associated lesions most prominent; often cyanotic with marked right-sided failure; big v and y descent; pulsation of liver	ECG — prominent P waves, or more often atrial fibrillation; RVH frequent; x-ray — enlarged right atrium and findings of associated lesions
Faint murmur of aortic insufficiency frequent; often ejection sound; A_2 often obscured and late; presystolic gallop frequent; protodiastolic with heart failure	Dyspnea, other symptoms of left heart failure; exertional syncope and angina frequent; arterial pulses soft and slowly rising	ECG — LVH often with marked ST-T abnormalities; x-ray — cardiomegaly and dilatation of ascending aorta; valve calcification seen in most adults
Holosystolic murmur of associated mitral insufficiency often present; A_2 soft; often paradoxical splitting of 2nd sound; presystolic gallop	Exertional dyspnea and syncope; pulse rises rapidly, collapses, then secondary rise; may have prominent a wave in neck from associated right-sided lesion	Murmur decreased by methoxamine or phenylephrine, increased by amyl nitrite or isoproterenol; ECG — LVH at times, Q waves suggestive of infarction; x-ray — cardiomegaly without poststenotic dilatation of aorta
Left diastolic blowing murmur at times; ejection sound does not occur; presystolic gallop at times	Characteristic facies; strabismus common; slow, small arterial pulse; blood pressure often higher in right arm than in left	ECG — LVH often with marked ST-T changes; x-ray — cardiomegaly with no poststenotic dilatation; often signs of idiopathic hypercalcemia in infancy
Left parasternal heave; second sound split, with late, usually soft P_2; ejection sound usually present; presystolic gallop	Distended neck veins, enlarged liver, edema with right heart failure; cyanosis at times when atrial septum is patent	ECG — shows RVH and right atrial hypertrophy or atrial fibrillation; x-ray — cardiomegaly and poststenotic dilatation of the pulmonary artery except in infundibular stenosis
Continuous murmur of patent ductus or bronchial collaterals may be heard; loud single second heart sound; no gallop	Cyanosis almost always; dyspnea, easy fatigability; retarded growth, squatting in childhood; clubbing; hypoxic spells at times	Murmur reduced by amyl nitrite; ECG — shows RVH with tall R in V_1; x-ray — normal size on PA; RVH on lateral; small pulmonary conus and decreased pulmonary vascularity
Ejection murmur of aortic stenosis in 10%; diastolic basal murmur of aortic insufficiency in 1/3; A_2 accentuated	Blood pressure elevated in arms, lower in legs; pulse late in legs; bruit can be heard and collaterals felt between ribs	ECG — high voltage QRS common, typical LVH at times; x-ray — cardiomegaly, after age 6 notching of ribs; indentation itself can be seen at times

(Continued on next page.)

DIFFERENTIAL DIAGNOSIS OF SYSTOLIC MURMUR (continued)

	Timing	Location	Radiation	Quality	Thrill
INTERVENTRICULAR SEPTAL DEFECT	Begins after 1st sound, may extend slightly past A_2	Along lower left sternal border	Widely; often to lower thorax posteriorly, not usually to neck	Medium frequency, of constant intensity	Frequent along lower left sternal border
INTERATRIAL SEPTAL DEFECT	Begins after 1st sound, ends well before 2nd sound	In 2nd or 3rd intercostal space along left sternal border	Usually localized; may radiate to apex	Harsh (crescendo-decrescendo) — early peak	No palpable thrill
RUPTURED PAPILLARY MUSCLE	Holosystolic begins with 1st sound; lasts to 2nd sound or just beyond	At apex	To axilla; to sternum at times with anterolateral muscle involvement	Loud, harsh, of constant intensity	Thrill infrequent
PAPILLARY MUSCLE DYSFUNCTION	Begins late in systole — persists almost to 2nd sound — holosystolic at times	At apex	To axilla	Often ejection (crescendo-decrescendo); constant intensity at times	Thrill infrequent
RUPTURED CHORDAE TENDINAE	Usually holosystolic; may begin late, always lasts to end of systole	At apex	Usually to axilla; at times to upper sternum	Harsh, constant intensity	Thrill infrequent
VENOUS HUM	Continuous, louder in diastole	Just above clavicle	2nd and 3rd interspaces	Humming, machinery-like	None
MAMMARY SOUFFLE	Continuous, louder in diastole	2nd intercostal space	None	Low frequency — not harsh	None
SUPRACLAVICULAR BRUIT	Begins after 1st sound, ends at 2nd sound	Supraclavicular space, usually on right	1st and 2nd interspaces	Ejection (crescendo-decrescendo); peak early or late	None
MUSICAL APICAL SYSTOLIC MURMUR	Usually holosystolic, affected by position and respiration	Apex	To left sternal border and base but not to axilla	Musical, single frequency	Occasional, at apex
PULMONARY EJECTION SYSTOLIC MURMUR	Begins after 1st sound, ends well before 2nd sound	2nd left intercostal space	None	Ejection (crescendo-decrescendo); peak prior to mid-systole	None

Associated Auscultatory Findings	Associated Signs and Symptoms	Key Laboratory Data
Mid-diastolic murmur at apex with large shunt; blowing murmur of pulmonic insufficiency with marked pulmonary hypertension; protodiastolic gallop at times	May be asymptomatic, or may show dyspnea, heart failure, impaired growth or even cyanosis	ECG — normal or LVH or biventricular hypertrophy; x-ray — enlargement of left ventricle, left atrium and pulmonary arteries; at times, right ventricle enlargement
Fixed splitting of second sound; diastolic rumble and presystolic gallop at times; apical holosystolic murmur when mitral valve is cleft	Left parasternal thrust, slender body build; signs of right heart failure and cyanosis in advanced disease	ECG — incomplete RBBB, complete in older; P-R prolongation in 20%, atrial fibrillation in older; x-ray — shows enlarged pulmonary arteries, cardiomegaly and small aortic knob
Protodiastolic gallop common	Severe dyspnea, signs of pulmonary edema, often shock	ECG — evidence of recent myocardial infarction
Gallop at times	May be evidence of increasing heart failure	ECG — evidence of infarct, old or recent
May have protodiastolic and/or presystolic gallop	Insidious or sudden increase in signs and symptoms of heart failure	Often other evidence of bacterial endocarditis — fever, anemia, blood cultures positive
Murmur loudest when sitting, louder when head turned away from side; obliterated by compression of cervical veins	None	None
Can be obliterated by local pressure	In last trimester of pregnancy and first few weeks postpartum	None
Other auscultatory findings normal	In children and young adults	None
Other auscultatory findings normal	Mostly in children	None
Decreased by sitting up	May be associated with increased flow states — fever, anemia, exercise, hyperthyroidism, etc.	Other evidence of cause of high output state, if one present

DIASTOLIC MURMUR

The presence of a murmur during cardiac diastole is one of the most valuable of all physical diagnostic signs since, except for a few innocent continuous vascular murmurs such as venous hum or mammary souffle, murmur in diastole always reflects heart disease.

Such murmurs are often more difficult to hear than systolic murmurs, and require various positions or maneuvers to be demonstrated. The effort is well worthwhile, since the murmur may reflect not only the more commonly occurring mitral stenosis, patent ductus arteriosus and aortic insufficiency, but a variety of other conditions.

The character of the murmur and the associated findings in a number of important causes of diastolic murmur are described separately in the following pages and then compared in tabular form.

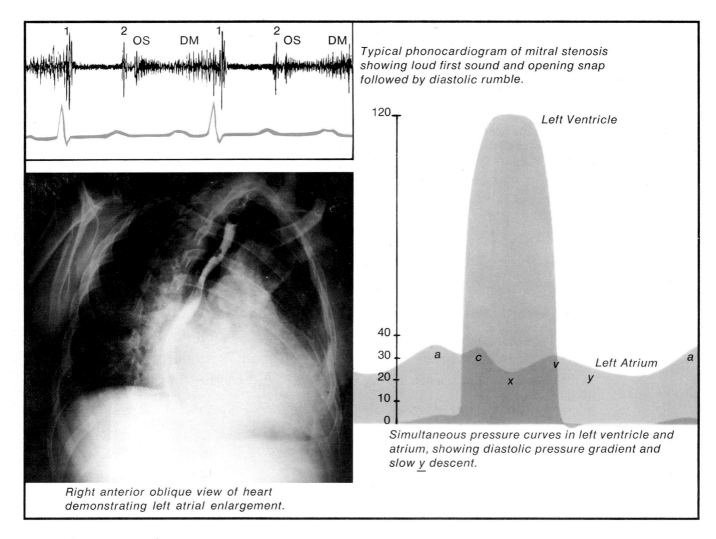

Typical phonocardiogram of mitral stenosis showing loud first sound and opening snap followed by diastolic rumble.

Simultaneous pressure curves in left ventricle and atrium, showing diastolic pressure gradient and slow *y* descent.

Right anterior oblique view of heart demonstrating left atrial enlargement.

Diastolic murmur due to
MITRAL STENOSIS

Mitral stenosis almost always results from rheumatic heart disease, although this cause can be documented in only about half the cases. Since it does not lead to clinically significant valve disease for 3 to 30 years after rheumatic fever (average 12 years), it is usually found in young adults, but can occur in the elderly; it is more common in women.

- Maximal at apex, often restricted to small area.
- Wide radiation uncommon; thrill at times, maximal at apex.
- Begins just after opening of mitral valve, about 0.1 second after A_2; may last almost to first sound.
- Low-frequency, "rumbling," decrescendo, with presystolic crescendo when rhythm is sinus; in atrial fibrillation, shorter and decrescendo; may be very faint when output is low.
- Louder after exercise, in left lateral supine position or after amyl nitrite; short apical systolic murmur common, even without mitral insufficiency.

ASSOCIATED FINDINGS

- Exertional dyspnea first symptom; orthopnea later.
- Cough, often nonproductive; or clear sputum, but frothy pink in marked pulmonary edema; hemoptysis of bright red blood may occur from rupture of dilated bronchial veins.
- Cyanosis, malar flush, icterus at times and reduced dyspnea — pulmonary hypertension and low output.
- Dysphagia, when present, due to enlarged atrium.
- Varied symptoms of systemic embolization common.
- Parasternal lift; normal or soft apical impulse.
- Loud first sound, loud P_2; opening snap of mitral valve about .08 second after A_2 at apex or along left sternal border, unless valve is immobile.
- Bilateral inspiratory moist rales; often wheezes from associated bronchitis.
- Signs of right heart failure occur late.
- ECG — QRS shows RVH or is normal; P mitrale when rhythm is sinus, but atrial fibrillation common.
- X-ray often shows pulmonary vascular congestion, enlarged pulmonary arteries and double density of enlarged left atrium.

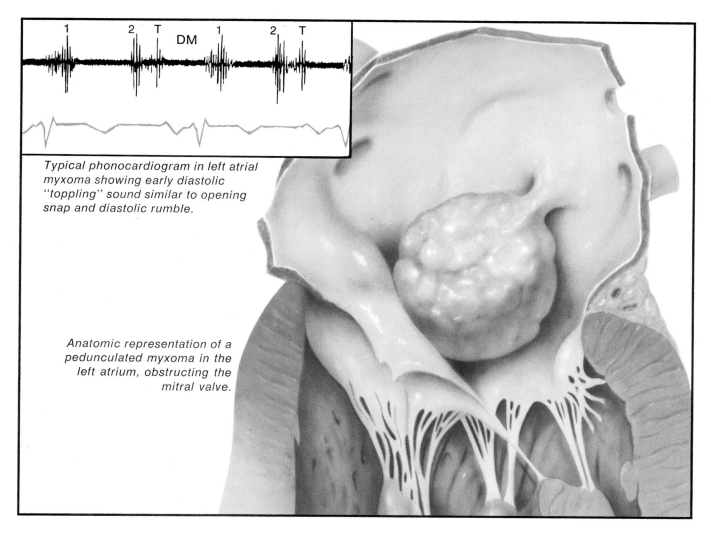

Typical phonocardiogram in left atrial myxoma showing early diastolic "toppling" sound similar to opening snap and diastolic rumble.

Anatomic representation of a pedunculated myxoma in the left atrium, obstructing the mitral valve.

Diastolic murmur due to
LEFT ATRIAL MYXOMA

Atrial myxoma is the most common intracavitary tumor of the heart. When on the right, it may simulate tricuspid stenosis or constrictive pericarditis; but it is more common on the left, where it mimics mitral stenosis. The sudden appearance of mitral stenosis in an older patient should arouse suspicion.

- Maximal at apex, with little radiation; thrill uncommon.
- Begins after opening of mitral valve, later than murmur of mitral stenosis; may last to 1st sound.
- Low-frequency, "rumbling," decrescendo, with presystolic crescendo when rhythm is sinus; murmur may alter with change of position; may be loudest when patient is sitting up.
- Short systolic murmur may be present.

ASSOCIATED FINDINGS

- Exertional dyspnea common.
- Dyspnea may be paroxysmal, associated with change of position; often most marked when sitting up.
- Recurrent acute episodes — syncope, pulseless episodes; transient hypotension, pulmonary edema.
- Epileptiform fits and coma may occur.
- Symptoms sometimes relieved by leaning forward.
- Varied symptoms from systemic embolization.
- Rapid deterioration is characteristic.
- Gangrene of nose and toes at times.
- S_2 often widely split because of widened systole.
- Sound similar to opening snap may occur, a little later than true opening snap.
- Atrial arrhythmias of various kinds and bundle branch block are common.
- Anemia, fever, increased ESR, weight loss and clubbing — nonspecific responses to tumor.
- ECG — normal or RVH; P mitrale may develop.
- X-ray will show relatively small left atrium — no calcification of mitral valve.
- Definite diagnosis by angiocardiography.

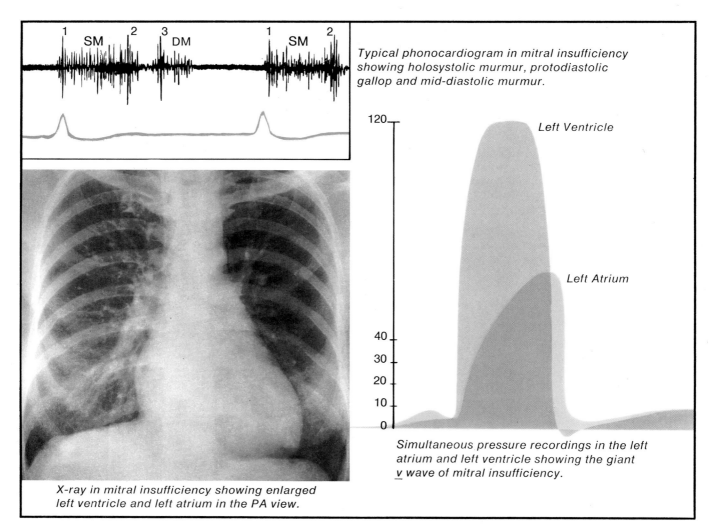

Typical phonocardiogram in mitral insufficiency showing holosystolic murmur, protodiastolic gallop and mid-diastolic murmur.

Simultaneous pressure recordings in the left atrium and left ventricle showing the giant *v* wave of mitral insufficiency.

X-ray in mitral insufficiency showing enlarged left ventricle and left atrium in the PA view.

Diastolic Murmur due to MITRAL INSUFFICIENCY

The most common cause for a diastolic apical murmur in mitral insufficiency is associated mitral stenosis, with both valve changes resulting from rheumatic carditis. It is important that a diastolic rumble can occur in this condition without any organic stenosis, probably because of greatly increased flow across the mitral valve, and can thus be heard in mitral insufficiency of *any* cause.

- Maximal at cardiac apex.
- Usually quite localized.
- Diastolic thrill not to be expected.
- Mid-diastolic . . . begins after third heart sound, during rapid filling period . . . later than murmur of organic mitral stenosis.
- Ends well before first sound.
- Decrescendo "rumbling" low-frequency murmur; no presystolic accentuation.
- Murmur always begins during early apical chest wall lift, from rapid filling.
- Holosystolic apical and axillary murmur *always* present.

ASSOCIATED FINDINGS

- Dyspnea, orthopnea, paroxysmal nocturnal dyspnea — with heart failure; may be preceded by fatigue and palpitation.
- Little water hammer pulse, rapid rise and fall.
- Vigorous apical impulse.
- First sound normal or loud, but often obscured by murmur; diminished with advanced failure.
- Protodiastolic gallop, often close to second sound — with or without heart failure.
- Presystolic gallop usually only from right heart . . . with pulmonary hypertension.
- Opening snap, when anterior leaflet mobile.
- Atrial fibrillation frequent.
- ECG: often LVH, occasionally only RVH; often nondescript; P mitrale when rhythm is sinus.
- X-ray: enlargement of left ventricle and atrium; at times of pulmonary artery and right ventricle.
- When a "rumbling" diastolic murmur occurs at the apex during acute rheumatic fever, it is accompanied by a holosystolic murmur and reflects transient insufficiency rather than stenosis.

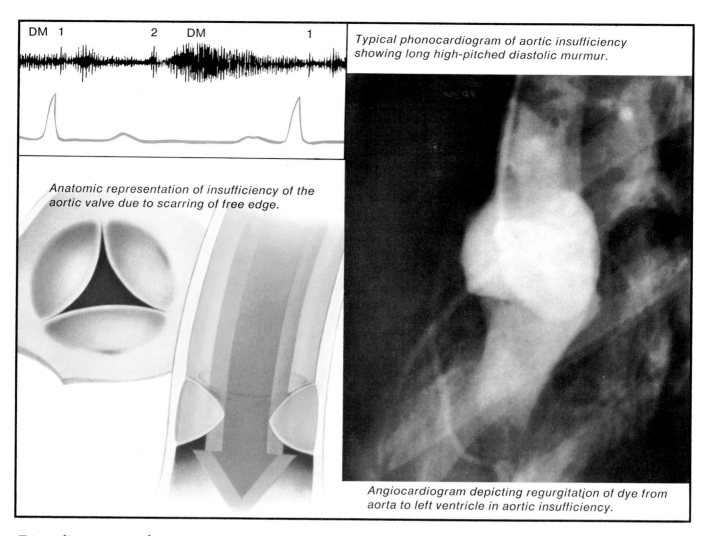

Typical phonocardiogram of aortic insufficiency showing long high-pitched diastolic murmur.

Anatomic representation of insufficiency of the aortic valve due to scarring of free edge.

Angiocardiogram depicting regurgitation of dye from aorta to left ventricle in aortic insufficiency.

Diastolic murmur due to AORTIC INSUFFICIENCY OF RHEUMATIC HEART DISEASE

The most common cause of diastolic murmur at the base of the heart is rheumatic aortic insufficiency. When the murmur is associated with the hemodynamic consequences of free aortic regurgitation and a history of rheumatic fever, diagnosis is not difficult; but when either of these is absent, it may be obscure.

- Maximal at 2nd ICS to right of sternum, or 3rd and 4th interspaces to left of sternum; radiates to apex, only slightly to the neck.
- Thrill uncommon; but if present at left sternal border and with musical "seagull" overtone, bacterial endocarditis should be suspected.
- Begins immediately after A_2 and may last through all or part of diastole.
- Decrescendo, high-pitched, lower-pitched at apex.
- Short systolic ejection murmur, often at 2nd interspace at right sternal border; occasionally with thrill.
- Diastolic rumble at apex, beginning after 3rd sound, often present (Austin Flint murmur); at times holosystolic harsh murmur at apex, due to functional mitral insufficiency.
- Murmurs of AI and Austin Flint decreased by amyl nitrite, unlike mitral stenosis.
- Murmur best heard when patient leans forward.

ASSOCIATED FINDINGS

- History of rheumatic fever in majority of cases.
- Palpitation and awareness of pulse in neck may be early symptoms; excessive sweating, dyspnea, paroxysmal nocturnal dyspnea, and finally orthopnea and signs of left and right heart failure.
- Exertional syncope and angina at times — but less than in aortic stenosis.
- Apical impulse forceful and diffuse.
- A_2 normal, or loud and ringing; first sound accentuated at times, often followed by ejection sound; protodiastolic gallop at apex.
- Pulse is wide and bounding — water-hammer; with associated stenosis may be double-peaked — "pulsus bisferiens."
- P-R interval often prolonged, fibrillation uncommon; x-ray shows cardiomegaly, dilated aorta.

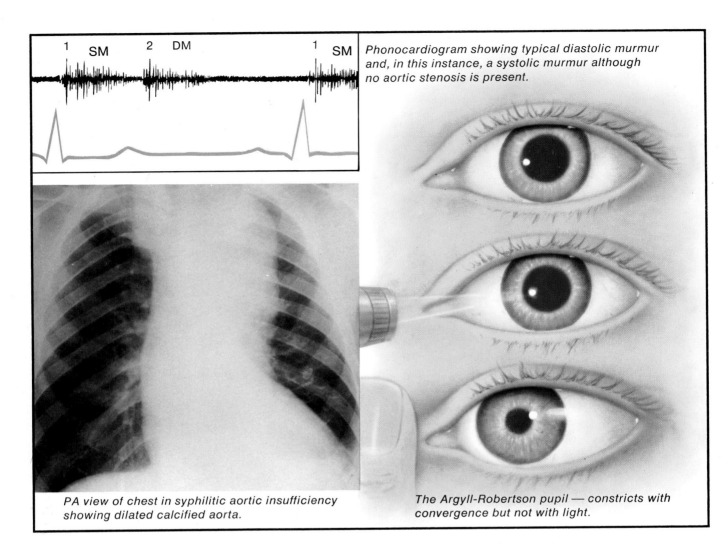

Phonocardiogram showing typical diastolic murmur and, in this instance, a systolic murmur although no aortic stenosis is present.

PA view of chest in syphilitic aortic insufficiency showing dilated calcified aorta.

The Argyll-Robertson pupil — constricts with convergence but not with light.

Diastolic murmur due to
AORTIC INSUFFICIENCY OF SYPHILITIC HEART DISEASE

The appearance of a basal diastolic murmur in an older person, especially a male, should suggest the possibility of syphilitic aortic insufficiency. Syphilitic aortitis, the cause of aortic aneurysm, will occur in 80% of untreated cases and result in clinically detectable lesions in 10%.

- Maximal at 2nd interspace to right of sternum; radiates down left sternal border to apex; if radiation is prominent down right sternal border, aneurysm or markedly dilated aorta may be present.
- Thrill uncommon; but when accompanied by musical "seagull" overtone, retroversion of leaflet may be present.
- Begins immediately after A_2, may last through all or part of diastole.
- Decrescendo; high-pitched, blowing, lower-pitched at apex.
- Short ejection murmur often at 2nd ICS at right sternal border, although stenosis never occurs; often a diastolic rumble at apex, beginning after 3rd sound (Austin Flint murmur).
- Murmur heard best when patient leans forward.

ASSOCIATED FINDINGS

- Palpitation and awareness of pulse in neck may be early symptoms; excessive sweating and intolerance to heat.
- Exertional syncope at times; exertional angina frequent, due to involvement of coronary ostia in about 20%; exertional dyspnea, paroxysmal nocturnal dyspnea — signs of left and right heart failure.
- Apical impulse forceful and diffuse.
- A_2 normal or loud and ringing; 1st sound accentuated at times; protodiastolic gallop at apex.
- Signs of CNS syphilis in more than a third — Argyll-Robertson pupil, diminished knee jerk, positive Romberg test; serology positive in 80%; specific immobilization in 100%; cerebrospinal titers positive in 50%.
- ECG shows LVH; atrial fibrillation uncommon.
- X-ray shows cardiomegaly and dilated aorta; may show calcification of ascending aorta — distinct aneurysm of thoracic aorta in less than 10%.

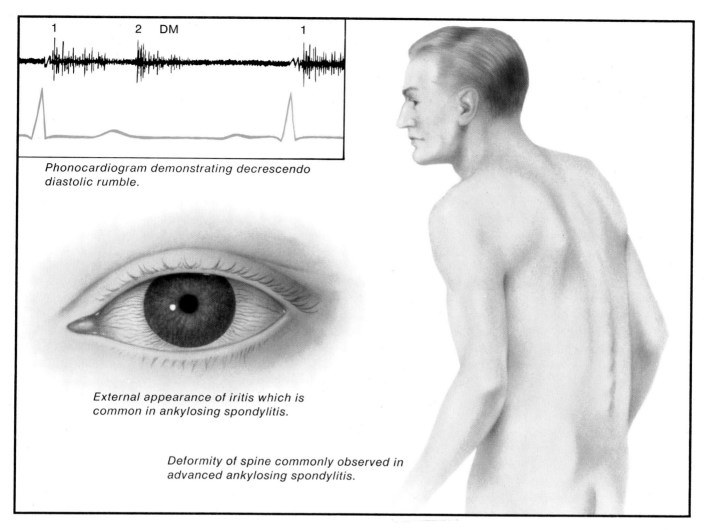

Phonocardiogram demonstrating decrescendo diastolic rumble.

External appearance of iritis which is common in ankylosing spondylitis.

Deformity of spine commonly observed in advanced ankylosing spondylitis.

Diastolic murmur due to AORTIC INSUFFICIENCY OF RHEUMATOID ARTHRITIS

Rheumatoid arthritis can affect the heart in several ways. Acute pericarditis can lead to adhesive pericarditis that never constricts; rheumatoid nodules occasionally occur in the heart. Clinically significant insufficiency of the aortic valve occurs in 10% of cases of ankylosing spondylitis.

- Maximal at 2nd interspace to right of sternum, or 3rd and 4th interspaces to left of sternum; radiates to apex, little to neck; thrill uncommon.
- Begins immediately after A_2, may last through all or part of diastole.
- Decrescendo — high-pitched and blowing.
- Short ejection murmur at aortic area; organic aortic stenosis absent.
- As in other types of aortic insufficiency, functional murmur of mitral stenosis (Austin Flint); mitral and tricuspid insufficiency may occur.
- Heard best when patient leans forward.

ASSOCIATED FINDINGS

- Most common in males; apparent 10 to 15 years after onset of arthritis.
- More likely in cases with other systemic signs.
- Palpitation, awareness of pulse in neck early; excessive sweating and heat intolerance reflect free regurgitation.
- Exertional dyspnea, paroxysmal nocturnal dyspnea, orthopnea and signs of left and right heart failure; exertional syncope and angina at times.
- Apical impulse forceful and diffuse; protodiastolic gallop often present.
- A_2 loud and ringing or normal; first sound often accentuated and followed by ejection sound.
- Pulse is wide and bounding — water-hammer.
- Spine kyphotic and rigid — no lumbar curve.
- Evidence of acute peripheral arthritis may be present, but without chronic changes.
- Latex fixation test for RA factor usually negative in cases of ankylosing spondylitis.
- ECG may show LVH; atrial fibrillation uncommon.
- X-ray: cardiomegaly, dilated aorta, "bamboo spine" of ankylosing spondylitis.

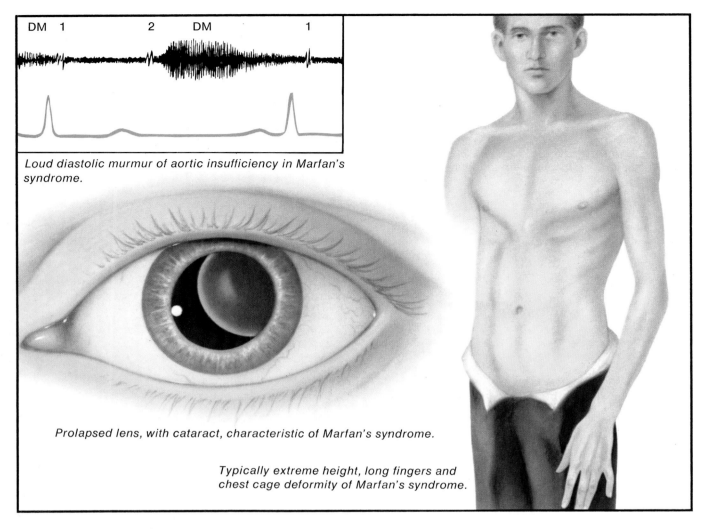

Loud diastolic murmur of aortic insufficiency in Marfan's syndrome.

Prolapsed lens, with cataract, characteristic of Marfan's syndrome.

Typically extreme height, long fingers and chest cage deformity of Marfan's syndrome.

Diastolic murmur due to AORTIC INSUFFICIENCY OF MARFAN'S DISEASE

Marfan's disease, an hereditary connective tissue disorder, may cause aortic insufficiency in two ways: chronically by dilatation of the aortic root, and acutely when dissecting aneurysm of the aorta suddenly distends the root. When the typical habitus of Marfan's disease is present, diagnosis is simple; the condition, however, can exist as a "forme fruste" with only aortic abnormalities.

- Maximal at the 2nd ICS to right of sternum; radiation down right sternal border.
- Begins immediately after A_2, may last through all or part of diastole.
- Decrescendo — high-pitched and blowing.
- Short systolic ejection murmur in aortic area; functional murmur of mitral stenosis (Austin Flint); mitral and tricuspid insufficiency at times.

ASSOCIATED FINDINGS

- Family history of typical body build, sudden death.
- Palpitation, awareness of neck pulsation, dyspnea, orthopnea, paroxysmal nocturnal dyspnea, exertional syncope, angina and left and right heart failure.
- Dyspnea, weakness and rending chest pain may be only symptoms when dissection is the cause — followed by death from cardiac tamponade.
- Apical impulse forceful and diffuse; protodiastolic gallop often present.
- A_2 normal or loud and ringing; 1st sound increased at times, often followed by ejection sound.
- Pulse wide and bounding — water-hammer type.
- Extreme height is common; long fingers and hyperextensible joints; high, narrow-arched palate; webbed fingers at times.
- Scoliosis, pectus carinatum and excavatum, genu valgum and recurvatum occasionally; prolapsed lenses and cataracts at times.
- When dissection is present: hypotension, absent pulses, pulsation of right sternoclavicular joint.
- ECG: LVH when dilatation is chronic; low voltage with dissection and tamponade.
- X-ray: cardiomegaly and dilatation of aorta — aneurysm of sinus of Valsalva may be present.

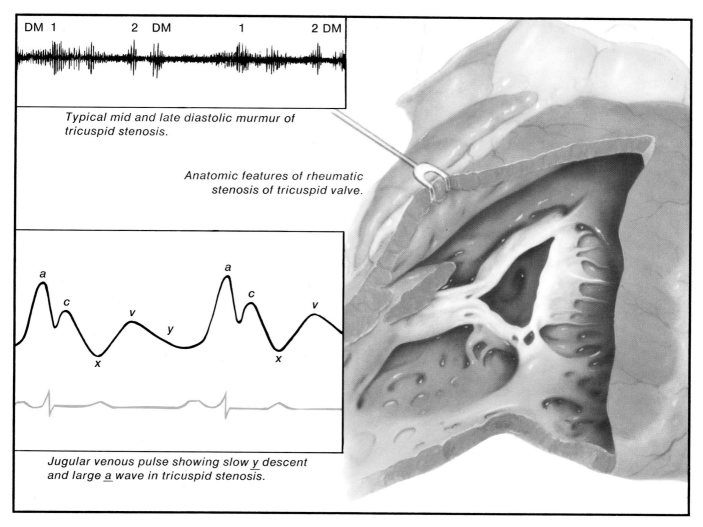

Typical mid and late diastolic murmur of tricuspid stenosis.

Anatomic features of rheumatic stenosis of tricuspid valve.

Jugular venous pulse showing slow y descent and large a wave in tricuspid stenosis.

Diastolic murmur due to
TRICUSPID STENOSIS

Isolated stenosis of the tricuspid valve — whether congenital, associated with metastatic carcinoid or rheumatic — is very rare. Together with other rheumatic heart lesions, it is not uncommon. It often complicates mitral stenosis and, since the murmurs of both are similar, the tricuspid stenosis may be easily overlooked.

- Maximal at 4th ICS to left of sternum; radiates toward apex and to xiphoid region.
- Thrill at times; may be increased by inspiration.
- Begins in early diastole after opening of tricuspid valve — usually hiatus before 1st sound.
- Decrescendo rumbling murmur with presystolic accentuation when sinus rhythm present; intensity augmented by inspiration.
- Somewhat higher-pitched than rumble of mitral stenosis; when fibrillation occurs, no presystolic accentuation — resembles aortic insufficiency.
- Murmurs of associated aortic and/or mitral disease usually present.

ASSOCIATED FINDINGS

- More common in women, usually young.
- Exertional but not paroxysmal nocturnal dyspnea; orthopnea less than in uncomplicated mitral stenosis.
- Hepatomegaly, ascites and edema, out of proportion to left-sided lesions.
- Symptoms may worsen acutely with paroxysmal atrial fibrillation.
- First sound not accentuated; P_2 not loud unless due to associated mitral stenosis.
- May have opening snap, usually loudest to left of sternum, occasionally to the right; opening snap earlier than usual snap of mitral stenosis — .05 to .07 second after A_2; if mitral valve is calcified and first sound soft, any opening snap heard is from tricuspid valve.
- Jugular venous pulse shows slow *y* descent; prominent *a* wave if rhythm is sinus.
- Presystolic pulsation of liver may be present.
- ECG shows right atrial hypertrophy if rhythm is sinus; usually atrial fibrillation.
- X-ray typically shows very large right atrium; also evidence of other valve lesions.

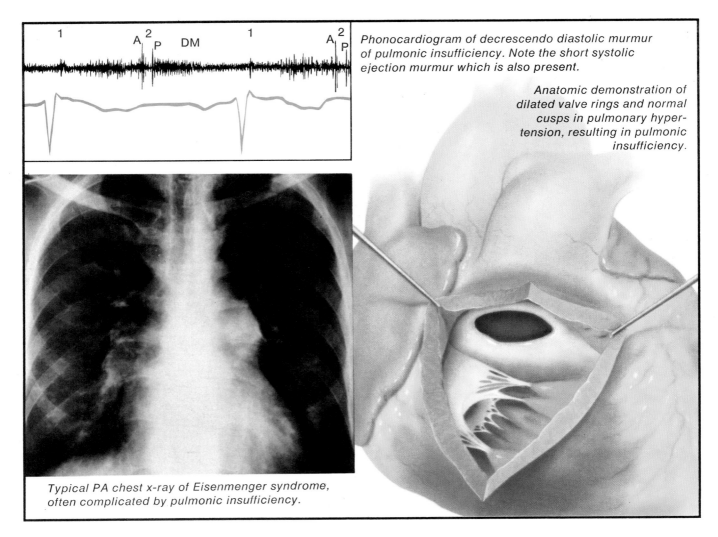

Phonocardiogram of decrescendo diastolic murmur of pulmonic insufficiency. Note the short systolic ejection murmur which is also present.

Anatomic demonstration of dilated valve rings and normal cusps in pulmonary hypertension, resulting in pulmonic insufficiency.

Typical PA chest x-ray of Eisenmenger syndrome, often complicated by pulmonic insufficiency.

Diastolic murmur due to
PULMONIC INSUFFICIENCY

Organic disease of the pulmonic valve causing insufficiency is quite rare; it may be rheumatic or due to congenital absence of valve, these lesions being well tolerated. Asymptomatic pulmonic insufficiency can also follow surgical repair of tetralogy of Fallot. Most commonly it is secondary to dilatation of the artery and valve ring due to increased pressure and/or flow.

- Maximal in 2nd ICS to left sternum; radiates along left sternal border, even to apex or aortic region at times; usually no thrill.
- Begins right after P_2, may last to first sound.
- Decrescendo, high frequency, blowing; made louder by inspiration; best heard when patient leans forward.
- Short systolic ejection murmur along left sternal border almost always present.
- Holosystolic murmur of tricuspid insufficiency, especially if rhythm is fibrillation.
- Murmur of mitral stenosis may be very faint when pulmonary hypertension is marked.

ASSOCIATED FINDINGS

- Dyspnea, breathlessness, weakness when due to pulmonary hypertension superimposed on a left-to-right shunt (Eisenmenger syndrome).
- Acrocyanosis and malar flush with mitral disease; cyanosis, clubbing and polycythemia with Eisenmenger syndrome.
- Parasternal heave due to RVH; systolic pulsation of pulmonary artery; palpable loud P_2.
- Prominent *a* wave in neck and presystolic gallop at apex if rhythm is sinus and ventricular septum is intact.
- Ejection sound often along left sternal border.
- Second sound single or finely split; wide splitting usually does not persist in atrial septal defect when Eisenmenger syndrome develops.
- Venous distention, hepatomegaly or edema with right heart failure.
- ECG: RVH; P mitrale with mitral disease; P pulmonale with Eisenmenger syndrome or atrial fibrillation.
- X-ray: large proximal pulmonary arteries, at times with calcification; clear distal lung fields.

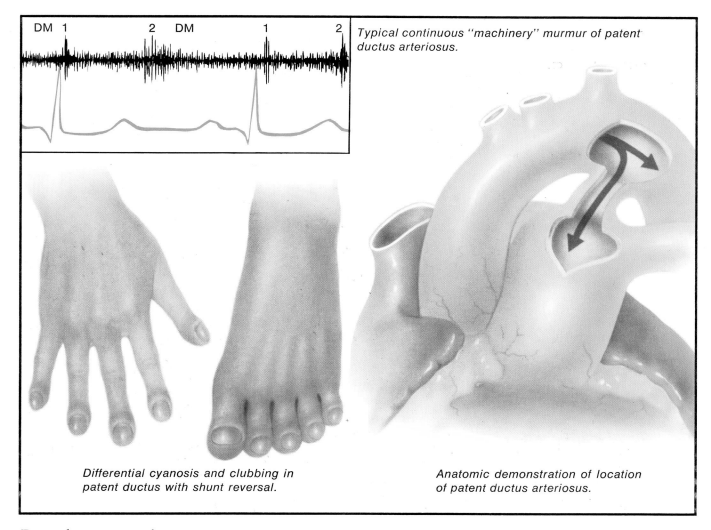

Typical continuous "machinery" murmur of patent ductus arteriosus.

Differential cyanosis and clubbing in patent ductus with shunt reversal.

Anatomic demonstration of location of patent ductus arteriosus.

Diastolic murmur due to PATENT DUCTUS ARTERIOSUS

The murmur of congenital patent ductus arteriosus is usually continuous and so characteristic that diagnosis can be made by auscultation alone. Since, however, it can be confused with the innocent venous hum and with other less common causes of a continuous murmur, attention must be given to the associated findings to insure correct diagnosis and appropriate surgery.

- Maximal in the 2nd ICS at left sternal border.
- Often radiates widely to left infraclavicular area, neck, and along the left sternal border.
- Thrill often present in the left second intercostal space and suprasternal notch.
- Usually continuous; if not, at end of systole and early diastole.
- Harsh "machinery" murmur, maximal at end systole.
- A separate holosystolic continuous murmur suggesting ventricular septal defect — or ejection murmur — is often heard along left sternal border.
- Mid-diastolic rumble of functional mitral stenosis is sometimes heard at the apex.
- Murmur often inapparent in first year of life or after Eisenmenger syndrome develops.

ASSOCIATED FINDINGS

- Growth and development usually normal.
- Recurrent pulmonary infections are common.
- Dyspnea and signs of left heart failure may develop where shunt is large.
- Pulse bounding; pulse pressure usually more than 45 mm Hg.
- Prominent and diffuse apical impulse.
- First sound may be accentuated; second sound may be paradoxically split with large shunt.
- P_2 may be loud with pulmonary hypertension.
- Protodiastolic gallop may be present at the apex.
- Cyanosis, clubbing may develop with reversal of shunt; greater in lower than in upper extremities.
- LVH, often with deep narrow Q wave and tall peaked T waves in V_5 and V_6; biventricular hypertrophy with pulmonary hypertension.
- X-ray: slight to moderate cardiomegaly, dilated aortic knob and large atrium; pulmonary arteries enlarged with increased pulmonary vasculature.

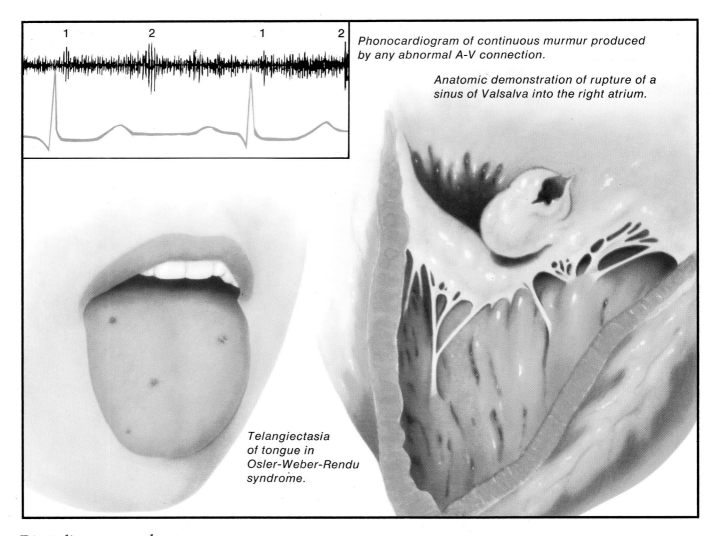

Phonocardiogram of continuous murmur produced by any abnormal A-V connection.

Anatomic demonstration of rupture of a sinus of Valsalva into the right atrium.

Telangiectasia of tongue in Osler-Weber-Rendu syndrome.

Diastolic murmur due to
EXTRACARDIAC A-V SHUNTING (OTHER THAN PATENT DUCTUS ARTERIOSUS)

At least nine conditions other than patent ductus can cause continuous murmur in both systole and diastole. All are uncommon and reflect abnormal connection between vessels outside ventricular cavities.

1. *Aorticopulmonary window*—a congenital abnormal opening between ascending aorta and main pulmonary artery. ▪ Mimics large patent ductus arteriosus; but murmur is maximal closer to sternum, often along LSB.

2. *Tetralogy of Fallot and pseudotruncus arteriosus* — both with very small or no pulmonary flow and extensive bronchial collaterals. ▪ Continuous murmur often maximal over back. ▪ Always cyanotic, anoxic spells. ▪ Parasternal lift of RVH. ▪ Loud single S_2.

3. *Truncus arteriosus* — single vessel leaves ventricles above ventricular septal defect; murmur due to flow from pulmonary arteries arising from trunk — often continuous, over back. ▪ Cyanosis, heart failure, anoxia. ▪ Parasternal lift, loud single 2nd sound; may have wide pulse pressure; systolic murmur along LSB at times. ▪ X-ray: often high left pulmonary artery.

4. *Systemic A-V aneurysm* — in chest wall (an A-V fistula, congenital or traumatic); if large, can lead to high output. ▪ Continuous murmur with systolic accentuation often maximal over palpable mass. ▪ Compression of A-V fistula will slow heart rate.

5. *Pulmonary A-V fistula* — usually congenital; often with other signs of Osler-Weber-Rendu syndrome — telangiectasia in mouth. ▪ Murmur often heard best in back. ▪ Usually some cyanosis. ▪ Density in lungs on fluoroscopy, decreases with Valsalva's maneuver.

6. *Sinus of Valsalva fistula* — Congenital aneurysm of sinus of Valsalva may rupture into right atrium or ventricle and produce acute failure. ▪ Continuous murmur at left sternal border. ▪ Protodiastolic gallop.

7. *Rupture of syphilitic aortic aneurysm into pulmonary artery* — Occurs with aneurysm of ascending aorta. ▪ Sudden appearance of continuous murmur along LSB. ▪ Right heart failure from associated compression of pulmonary artery, at times.

8. *Coronary A-V fistula* — Congenital; may enter right atrium or ventricle. ▪ Continuous murmur low along LSB. ▪ Murmur reduced in systole if shunt enters right ventricle. ▪ Often wide pulse pressure.

9. *Distal pulmonary artery coarctations* — May occur as isolated congenital defect or in association with other lesion, *e.g.,* supravalvular aortic stenosis. ▪ Continuous murmur may be maximal in back.

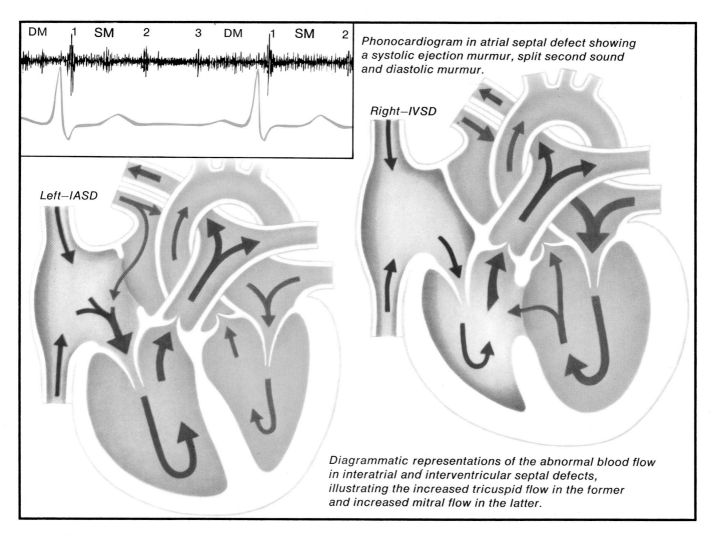

Phonocardiogram in atrial septal defect showing a systolic ejection murmur, split second sound and diastolic murmur.

Diagrammatic representations of the abnormal blood flow in interatrial and interventricular septal defects, illustrating the increased tricuspid flow in the former and increased mitral flow in the latter.

Diastolic murmur due to SEPTAL DEFECTS

The diastolic murmurs commonly produced by both atrial and ventricular septal defects are functional rather than organic, representing increased flow across the A-V valves. They are important in that, if misconstrued as mitral stenosis, the lesions may be considered more complicated than they actually are.

- Maximal at the lower left sternal border or inside apex in atrial septal defect; at the apex in ventricular septal defect.
- Well localized; does not radiate.
- Usually no thrill in diastole.
- Short, mid-diastolic; follows third sound when present; decrescendo, rumbling, no presystolic accentuation.
- Both conditions may also have diastolic blowing murmur along left sternal border with pulmonary hypertension; since flow is then reduced, no diastolic rumble in such cases.
- IVSD may have diastolic blowing murmur along LSB without pulmonary hypertension, because of prolapse of aortic or, rarely, pulmonary leaflet.
- Systolic murmurs of primary lesions are present.

ASSOCIATED FINDINGS

- Frequent respiratory infections.
- Exertional dyspnea, even heart failure.
- Vigorous apical impulse with IVSD; parasternal lift with IASD.
- Second sound has exaggerated split; moves with respiration in IVSD; wide fixed split in IASD.
- P_2 loud with pulmonary hypertension or very large shunt; no opening snap; presystolic apical gallop in IASD, protodiastolic gallop in IVSD; children with IASD may have third sound just prior to rumble; ejection sound in pulmonary artery at times.
- ECG: incomplete RBBB, often prolonged P-R with IASD; LVH or biventricular hypertrophy with IVSD large enough to produce rumble.
- X-ray: enlarged pulmonary arteries and increased pulmonary vasculature; globular cardiomegaly in IASD; definite left ventricular enlargement in IVSD with large shunt; atrial enlargement in both.

DIFFERENTIAL DIAGNOSIS OF DIASTOLIC MURMUR

	Location	Radiation	Thrill	Timing	Character
MITRAL STENOSIS	At apex, often quite localized	Minimal, to axilla at times	At times; maximal at apex	Begins 0.1 second after A_2; may last almost to 1st sound	Low frequency, rumbling — decrescendo; crescendo at S_4 with sinus rhythm
LEFT ATRIAL MYXOMA	At apex	Little or none	Uncommon	Begins later than in mitral stenosis; murmur may persist till first sound	Low frequency, rumbling — decrescendo; crescendo at S_4 with sinus rhythm
MITRAL INSUFFICIENCY	At apex	None	Uncommon	Mid-diastolic; begins after 3rd sound, ends before first sound	Rumbling decrescendo; no presystolic accentuation
RHEUMATIC AORTIC INSUFFICIENCY	At 2nd ICS to right of sternum, or 3rd & 4th ICS to left of sternum	To apex; only slightly to neck	Uncommon; at times with "seagull" murmur of associated bacterial endocarditis	Begins immediately after A_2; may last through all or part of diastole	Decrescendo, high-pitched, blowing
SYPHILITIC AORTIC INSUFFICIENCY	2nd interspace to right of sternum	Down LSB to apex; down right border with aneurysm or marked dilatation	Uncommon; may occur with "seagull" murmur of everted leaflet	Begins immediately after A_2 and may last through all or part of diastole	Decrescendo, high-pitched, blowing
RHEUMATOID AORTIC INSUFFICIENCY	2nd ICS to right of sternum or 3rd & 4th ICS to left	To apex along left sternal border, little to neck	Uncommon	Begins immediately after A_2 and may last through all or part of diastole	Decrescendo, high-pitched, blowing
MARFAN'S AORTIC INSUFFICIENCY	2nd ICS to right of sternum	Down right sternal border	Uncommon	Begins immediately after A_2; may last through all or part of diastole	Decrescendo, high-pitched, blowing
TRICUSPID STENOSIS	4th ICS to left of sternum	To apex & xiphoid region	At times; increased by inspiration	Begins in early diastole, ends before first sound	Decrescendo rumbling, presystolic accentuation with sinus rhythm; high-pitched

Other Cardiac Findings	Associated Signs and Symptoms	Key Laboratory Data
Louder in left lateral supine position, after exercise or amyl nitrite; often short apical systolic murmur, loud 1st sound and P_2; opening snap; parasternal lift in RVH	Exertional dyspnea, orthopnea; cough (often nonproductive) or frothy pink sputum, hemoptysis; cyanosis, malar flush; when severe, inspiratory rales, signs of RHF	ECG: QRS show RVH or normal; P mitrale or atrial fibrillation; x-ray: double density of enlarged left atrium, enlarged pulmonary arteries
Murmur may change with position, often loudest on sitting; 1st sound not accentuated, 2nd sound widely split at times, perhaps like opening snap	Exertional and sedentary dyspnea, syncope, pulseless episodes, epileptiform fits, gangrene of nose and toes; rapid deterioration; fever, weight loss, clubbing	ECG: atrial arrhythmias; BBB common; normal or RVH; P mitrale at times; x-ray: relatively small left atrium; no valve calcification; anemia, increased ESR
Loud apical holosystolic murmur always present; protodiastolic gallop; 1st sound loud, but often obscured by opening snap; vigorous apical impulse, often with diastolic expansion	Fatigue, palpitation, dyspnea, orthopnea, PND, left and right heart failure; little water hammer pulse	ECG: often LVH; occasionally only RVH; at times nondescript; P mitrale with sinus rhythm, atrial fibrillation common; x-ray: enlargement of LV&LA; at times RA&RV
Murmur heard best when patient leans forward; often short ejection systolic murmur at aortic area; diastolic rumble at apex beginning after 3rd sound at times; protodiastolic gallop; loud S_1, A_2	History of rheumatic fever; palpitation & awareness of pulse in neck; excessive sweating, dyspnea; then left & right failure; exertional syncope & angina; water hammer pulse, wide pulse pressure	ECG: left axis deviation; typical LVH in many; P-R often prolonged; atrial fibrillation uncommon; x-ray: cardiomegaly, dilated aorta
Heard best when patient leans forward; often short ejection murmur at aortic area; diastolic rumble at apex & A_2 loud at times; protodiastolic gallop at apex	Palpitation and awareness of pulse in neck, excessive sweating, exertional syncope, angina, left and right heart failure; bounding pulse; CNS syphilis in 1/3	Serologic test for syphilis often positive. ECG: LVH; fibrillation uncommon; x-ray: cardiomegaly and dilated aorta
Heard best when patient leans forward; often short ejection murmur at aortic area; diastolic rumble at apex & A_2 loud at times; protodiastolic gallop at apex	Palpitation, awareness of pulse in neck, excessive sweating, exertional syncope, angina, dyspnea, orthopnea, heart failure; bounding pulse; kyphosis, no lumbar curve; iritis at times	Latex fixation for rheumatoid factor usually negative. ECG: LVH; AF uncommon; x-ray: cardiomegaly, dilated aorta, "bamboo spine" of ankylosing spondylitis
Heard best when patient leans forward; often short ejection murmur at aortic area; diastolic rumble at apex & loud A_2 at times; protodiastolic gallop at apex	Limbs elongated, "double-jointed"; arched palate, prolapsed lenses, pectus carinatum or excavatum; palpitation, exertional syncope and angina, heart failure, wide bounding pulse	ECG: LVH; AF uncommon; x-ray: cardiomegaly and dilated aorta; aneurysm of sinus of Valsalva at times
Murmurs of associated aortic & mitral disease usually present; early opening snap; P_2 not loud unless mitral stenosis also present	Exertional dyspnea, signs of right heart failure; slow y descent, often big a wave in jugular venous pulse; presystolic pulsation of liver may be present	ECG: right atrial hypertrophy or atrial fibrillation; x-ray: right atrial enlargement & evidence of other valve lesions

(Continued on next page.)

DIFFERENTIAL DIAGNOSIS OF DIASTOLIC MURMUR (continued)

	Location	Radiation	Thrill	Timing	Character
PULMONIC INSUFFICIENCY	2nd ICS to left of sternum	Along LSB even to apex; to aortic region at times	Usually none	Begins right after P_2; may last to first sound	Decrescendo, high frequency, blowing
PATENT DUCTUS ARTERIOSUS	2nd ICS at left sternal border	Wide; to left infraclavicular area, neck, along LSB	Often; in 2nd left ICS and suprasternal notch	Usually continuous; if not, at end of systole and in early diastole	Rumbling machinery murmur, loudest in late systole
AORTICO-PULMONARY WINDOW	Near sternum, along LSB	Wide	Often, along lower LSB	Continuous	Machinery, peak at end of systole
TETRALOGY OF FALLOT & PSEUDOTRUNCUS	Often over upper back	To entire heart and lungs	None	Continuous	Machinery, peak at end of systole
TRUNCUS ARTERIOSUS	Often over upper back	To entire heart and lungs	None	Continuous	Machinery, peak at end of systole
SYSTEMIC A-V ANEURYSM	Maximal over palpable chest mass	Depends on primary location	Common, over palpable mass	Continuous	Machinery, peak at end of systole
PULMONARY A-V FISTULA	At lung lesion; often in back	Depends on location of fistula	Uncommon	Continuous	Machinery, peak at end of systole
SINUS OF VALSALVA FISTULA	At left sternal border	Often wide	Common, at left sternal border	Continuous	Machinery, peak at end of systole
RUPTURE OF LUETIC ANEURYSM INTO PULMONARY ARTERY	At left sternal border	Often wide	May occur, at LSB	Continuous	Machinery, peak at end of systole
CORONARY A-V FISTULA	Lower left sternal border	Often wide	May occur	Continuous	Machinery, peak at end of systole
DISTAL PULMONARY ARTERY COARCTATIONS	Maximal in back	Often wide	Usually none	Continuous	Machinery, peak at end of systole
ATRIAL SEPTAL DEFECT	Lower left sternal border or inside apex	Localized; no radiation	None	Short, mid-diastolic	Decrescendo, rumbling
VENTRICULAR SEPTAL DEFECT	At apex	Localized; no radiation	None	Short, mid-diastolic; follows third sound when present	Decrescendo, rumbling

Other Cardiac Findings	Associated Signs and Symptoms	Key Laboratory Data
Murmur heard best when patient leans forward; louder on inspiration, short ejection systolic murmur along LSB usually present; loud, single or finely split P_2; parasternal lift	Dyspnea, weakness; acrocyanosis & flush in mitral disease; cyanosis, clubbing & polycythemia in Eisenmenger syndrome; prominent a wave in neck and presystolic gallop if sinus rhythm normal & septa intact; often ejection sound; RHF	ECG: RVH; P mitrale with mitral disease; P pulmonale with Eisenmenger syndrome or atrial fibrillation; x-ray: large proximal pulmonary arteries
Often ejection murmur along LSB; holosystolic at times; diastolic rumble of functional mitral stenosis at times, S_1 and P_2 loud; 2nd sound may be paradoxically split	Development usually normal; recurrent pulmonary infections; signs of left failure; bounding pulse; cyanosis with reversal of shunt — greater in lower than upper extremities	ECG: LVH often with deep Q wave, V_5, V_6 — biventricular hypertrophy with pulmonary hypertension
Like patent ductus arteriosus	Like patent ductus arteriosus; apt to be severe	Like patent ductus arteriosus
Parasternal lift of RVH; loud single second sound	Cyanotic, anoxic spells	ECG: RVH; x-ray: usually small pulmonary artery
Parasternal lift, loud 2nd sound; wide pulse pressure; systolic murmur along LSB at times	Cyanosis, anoxic spells, heart failure at times	ECG: RVH; x-ray: high left pulmonary artery often visible
Compression of A-V fistula will slow heart rate	Wide pulse pressure and other signs of high output state	Cardiomegaly if output very high
—	Often multiple telangiectasia, e.g., in mouth; usually some cyanosis	Density seen in lungs, gets smaller with Valsalva's maneuver
Protodiastolic gallop	Often other signs of Marfan's syndrome; commonly acute heart failure	Enlarged sinus of Valsalva may be seen on x-ray
—	Right heart failure at times	X-ray: evidence of aneurysm of ascending aorta
Murmur reduced in systole if shunt enters right ventricle	Often wide pulse pressure	—
Often signs of supravalvular aortic stenosis	—	Definite diagnosis by angiography
Systolic ejection murmur along LSB; may have blowing diastolic murmur of pulmonic insufficiency, wide fixed split of 2nd sound, presystolic gallop	Frequent respiratory infections; exertional dyspnea; heart failure at times; parasternal lift	ECG: incomplete RBBB; often prolonged P-R; x-ray: globular cardiomegaly with enlarged pulmonary arteries & increased vasculature
Holosystolic murmurs along LSB; may have diastolic blowing murmur of pulmonic insufficiency with protodiastolic gallop; P_2 often loud, closely split	Frequent respiratory infections, exertional dyspnea, even heart failure; vigorous apical impulse	ECG: LVH or biventricular hypertrophy; x-ray: LVH; dilated pulmonary arteries and increased pulmonary vasculature

ABNORMAL HEART SOUNDS

The development of modern phonocardiography and the precise correlation of its findings with hemodynamic data derived from cardiac catheterization has resulted, in the last few decades, in a greatly increased appreciation of the diagnostic value of abnormalities of the first and second heart sounds. Much specific diagnostic information can now be obtained not only from the intensity but also from the timing of these sounds. This information, in conjunction with other cardiac findings, has made possible increasing precision in cardiac diagnosis.

In this chapter the cardiac findings in 20 important causes of abnormal heart sounds are presented individually and then reviewed in tabular form.

GENERATION OF HEART SOUNDS

The 1st and 2nd heart sounds, which normally mark the beginning and end of systole, undergo many alterations in various disease states. To understand the significance of these changes it is very helpful to understand the factors that contribute to *normal* heart sounds.

In the past there has been much debate as to whether the sounds are produced by the muscles, valves or other anatomic entities. This is no more easily answered than the question of whether the mouthpiece, the stops or the tubing produce the sound of a wind instrument — for all together produce vibration in a column of air that constitutes the sound. Similarly, the action of the muscles and valves of the heart results in the sudden acceleration and deceleration of a column of blood which, in fact, produces the sound.

Thus, the 1st heart sound, at the onset of systole, has three components. The first of these occurs during early systole when ventricular contraction, which has been compressing blood and pushing the A-V valves upward, pushes these valves to their limit and the blood is suddenly stopped, not moving again until the semilunar valves start to open. This sound is ordinarily not important; but in mitral stenosis, in which the valve is fixed and rigid, this deceleration is quite sharp and produces the loud 1st sound heard in that disease.

The second and major component of the 1st heart sound is produced by the sudden acceleration of blood as contraction continues and the semilunar valves open.

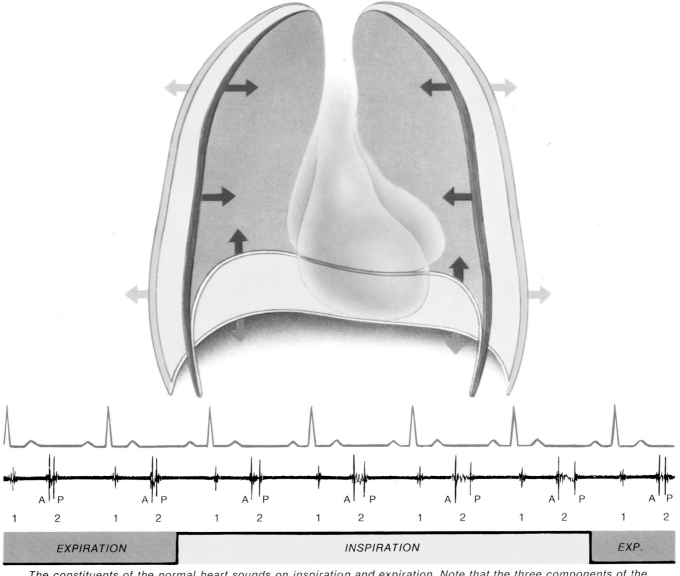

The constituents of the normal heart sounds on inspiration and expiration. Note that the three components of the first sounds are not distinguishable at ordinary recording speeds.

The intensity of this component is directly related to ventricular contractility: the more forceful the contraction, the louder the first sound. Even the soft 1st sound produced by P-R prolongation is due to decreased contractility — rather than to the position of the A-V valves. Indeed the A-V valves are often closed in normal subjects prior to the onset of the 1st sound.

The 3rd component of the 1st sound is probably produced in the aortic root, and is more of laboratory than clinical significance.

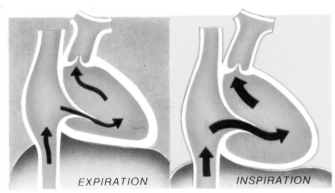

Inspiration, by decreasing intrathoracic pressure, increases venous return and right ventricular stroke volume.

The 2nd heart sound always has two components, both of which are produced by sudden changes in blood flow resulting from closure of the aortic and pulmonary valves. It has traditionally been said that the 2nd sound heard in the second interspace along the right sternal border reflects aortic closure, and along the left sternal border pulmonic closure. In fact, both closures can at times be heard in either position. The components of the 2nd sound can be identified only by hearing both and assessing their response to respiration. In general, the 2nd sound heard at the cardiac apex can be assumed to be aortic in origin since the pulmonic component rarely transmits that far.

Since the higher-pressure left ventricular systole is somewhat shorter than the systole on the right, the aortic component normally precedes the pulmonic. In expiration they may be so close that this separation will be difficult to appreciate, and at most they are a few hundredths of a second apart.

In inspiration, however, the increased venous return to the right ventricle further prolongs right ventricular systole and delays pulmonic closure. Then, splitting of the 2nd sound should be readily apparent; the separation may be as much as .05 of a second.

Exaggerated splitting can be produced by prolongation of right ventricular systole — hemodynamically as in pulmonic stenosis, pulmonary fibrosis or atrial septal defect, or electrically as in right bundle branch block.

When left ventricular systole is prolonged — hemodynamically as in aortic stenosis, severe left ventricular failure or patent ductus arteriosus, or electrically as in LBBB — the aortic closure will follow well after the pulmonic in expiration and splitting *will* occur. When, however, right ventricular systole is prolonged by the increased venous inflow of inspiration, the pulmonic closure will also come later and the splitting will disappear. Such a respiratory phenomenon is called paradoxical splitting.

The intensity of the aortic and pulmonic components is directly proportional to the pressure in the aorta and pulmonary artery, respectively, when they are recorded within the vascular system. Since the pulmonary artery root is much closer to the chest wall than the aortic root, the pulmonic component will seem louder at any given pressure than the aortic component.

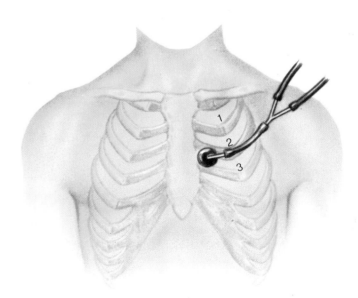

Location on chest wall where splitting of the second sound is best heard.

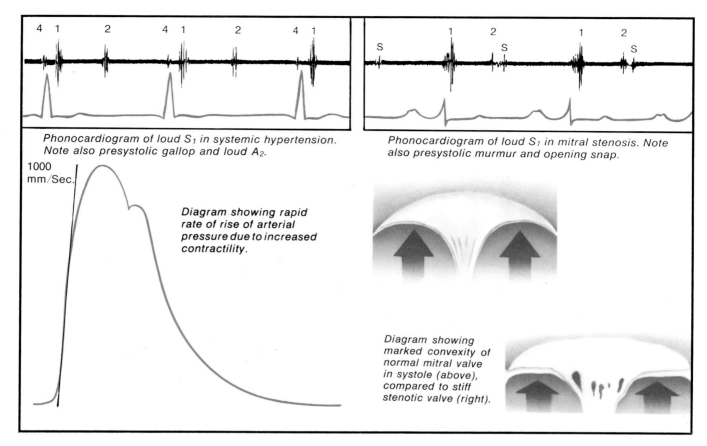

Phonocardiogram of loud S_1 in systemic hypertension. Note also presystolic gallop and loud A_2.

Phonocardiogram of loud S_1 in mitral stenosis. Note also presystolic murmur and opening snap.

Diagram showing rapid rate of rise of arterial pressure due to increased contractility.

Diagram showing marked convexity of normal mitral valve in systole (above), compared to stiff stenotic valve (right).

Loud S_1 due to INCREASED CONTRACTILITY

Contractility is increased whenever the ventricle must pump more blood — as in fever, anemia, aortic insufficiency or mitral insufficiency; or expel blood against increased resistance — as in systemic hypertension; or contract forcefully in response to endogenous or exogenous catecholamines, *e.g.*, with rapid heart rate or epinephrine or isoproterenol infusion. A loud 1st heart sound should be expected in all these conditions.

■ Loud 1st sound heard best at apex. ■ A_2 loud in hypertension and often in aortic insufficiency.
■ Apical presystolic gallop in fever, anemia, systemic hypertension at times. ■ Apical protodiastolic gallop in aortic and mitral insufficiency; in all when CHF supervenes. ■ Diastolic blowing murmur at base of heart in aortic insufficiency. ■ Holosystolic harsh murmur at apex in mitral insufficiency, often obscures loud 1st sound. ■ Short ejection systolic murmur along LSB in fever and anemia; at RSB in aortic insufficiency.

ASSOCIATED FINDINGS

■ Pulse vigorous with steep upstroke in all; steep collapse in aortic insufficiency — to some extent in anemia, fever, mitral insufficiency. ■ Wide pulse pressure. ■ Vigorous apical impulse. ■ ECG: high voltage QRS can occur in all; definite LVH in hypertension, aortic and mitral insufficiency. ■ X-ray: cardiomegaly in all if severe; left atrial enlargement in mitral insufficiency.

Loud S_1 due to MITRAL STENOSIS

Increased intensity of the 1st heart sound is one of the most consistent and readily appreciated findings in mitral stenosis. The discovery of an unexpected loud 1st sound should always prompt a careful search for opening snap and diastolic apical rumble, which includes listening with the patient in the left lateral supine position.

■ Loud 1st sound, best heard at apex; often sharp, snapping quality. ■ P_2 often accentuated. ■ Normal inspiratory splitting reduced at times. ■ Early diastolic opening snap often transmitted from apex to base — simulates split 2nd sound. ■ Decrescendo rumbling murmur, localized at apex; begins after opening snap, lasts almost to 1st sound; presystolic accentuation if rhythm is sinus. ■ Associated apical holosystolic murmur of mitral insufficiency often present.
■ Presystolic gallop along LSB at times.

ASSOCIATED FINDINGS

■ History of rheumatic fever in about half.
■ Progressive exertional dyspnea and orthopnea.
■ Nonproductive cough, hemoptysis. ■ Signs of pulmonary edema and right heart failure at times.
■ Localized apical impulse; may have parasternal lift. ■ ECG: P mitrale or atrial fibrillation; right ventricular hypertrophy at times. ■ X-ray: enlarged left atrium, pulmonary arteries; increased pulmonary vasculature when severe.

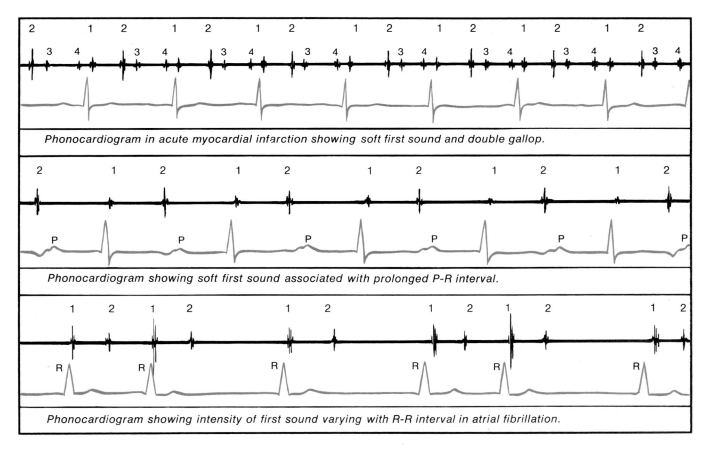

Phonocardiogram in acute myocardial infarction showing soft first sound and double gallop.

Phonocardiogram showing soft first sound associated with prolonged P-R interval.

Phonocardiogram showing intensity of first sound varying with R-R interval in atrial fibrillation.

Soft S_1 due to
DECREASED CONTRACTILITY

Since a soft first sound occurs in many forms of left ventricular dysfunction, it alone is not a diagnostic finding of great strength. Serial changes in a single patient may be important clues in following the evolution of myocardial infarction or digitalis intoxication. (The soft first sound of P-R prolongation is also due to decreased contractility.)

- First sound soft and often split wherever it is heard.
- Paradoxical split of second may also be present when left ventricular function is markedly depressed.
- Pulmonic component of second sound often accentuated with heart failure. ■ Gallop rhythm, especially presystolic, is frequent. ■ Presence of murmurs depends on underlying disease.

ASSOCIATED FINDINGS

- Often follows the chest pain, dyspnea and sweating of acute myocardial infarction. ■ History of long-standing valvular disease or cardiomyopathy may be present. ■ Pulse pressure may be narrow. ■ Pulsus alternans often present. ■ ECG: often evidence of recent infarct and/or prolonged P-R. ■ May show LVH when cardiomyopathy is the cause of decreased contractility. ■ Chest x-ray: cardiomegaly frequently present.

Varying intensity of S_1 due to
ATRIAL FIBRILLATION

The most characteristic physical finding in atrial fibrillation is total irregularity of the ventricular rate. These irregularities lead to a very different duration of the diastolic filling period, and thus to beat-to-beat variation in diastolic stretch and contractility. Fibrillation can occur in any form of heart disease or without heart disease, but is most common in arteriosclerotic and rheumatic heart disease and in thyrotoxicosis.

- First sound varies from soft to loud — generally loud following short R-R intervals and soft following long R-R intervals. ■ Character of murmur and other heart sounds dependent on underlying disease.
- Presystolic gallop and *a* wave in neck are not present, but protodiastolic gallop can occur.

ASSOCIATED FINDINGS

- Breathlessness and awareness of palpitation will usually be present. ■ Other symptoms dependent on underlying disease. ■ Amplitude of pulse and systolic blood pressure will also vary beat to beat. ■ ECG: totally irregular QRS rhythm with continuous atrial activity that is variable in rate and form — seen best in V_1. ■ X-ray: dependent on underlying disease; may be normal.

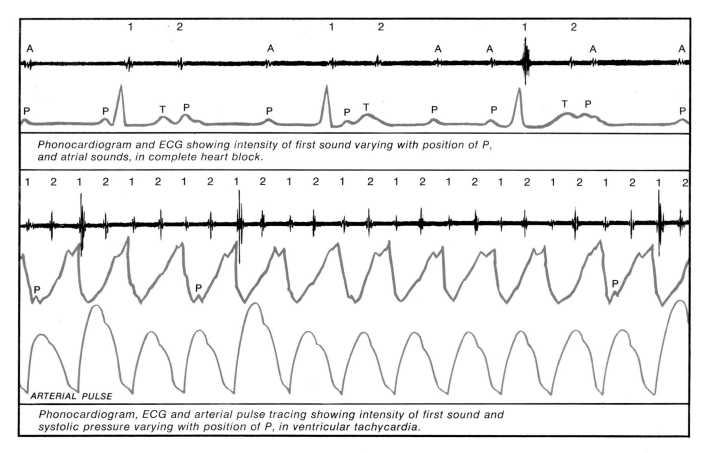

Phonocardiogram and ECG showing intensity of first sound varying with position of P, and atrial sounds, in complete heart block.

Phonocardiogram, ECG and arterial pulse tracing showing intensity of first sound and systolic pressure varying with position of P, in ventricular tachycardia.

Varying intensity of S_1 due to
COMPLETE HEART BLOCK

Complete heart block can result from acute myocardial infarction, digitalis intoxication, rheumatic fever and many other causes, but is most common in the elderly — due to patchy fibrosis of or around the conducting system. The varying intensity of the first sound is due to varying timing of atrial contraction.

■ First sound varies in intensity, even though the ventricular rate is regular and usually very slow.
■ Variation appreciated best at apex, but can be heard all over the heart. ■ Second sound split variably, depending on site of idioventricular pacemaker.
■ Intensity of components of second sound dependent on underlying disease. ■ Atrial sounds heard at variable times in diastole — best along LSB. ■ Ejection murmur often present because of slow rate and large stroke volume.

ASSOCIATED FINDINGS

■ Intermittent syncope common — Stokes-Adams syndrome. ■ Pulse full — pulse pressure wide.
■ Cannon *a* waves in neck, when atrial contraction occurs during ventricular systole. ■ ECG: regular, slow QRS, the rate of which is unrelated to atrial rate; QRS may be normal or aberrant in form. ■ X-ray: occasionally shows calcification in ventricular septum.

Varying intensity of S_1 due to
VENTRICULAR TACHYCARDIA

Ventricular tachycardia, a life-threatening arrhythmia, may complicate myocardial infarction, digitalis intoxication or, less frequently, other forms of heart disease. Identification of appropriate physical signs can lead to early diagnosis, treatment and survival.

■ First sound varies in intensity despite basically regular and rapid rate, because of varying timing of atrial contraction. ■ Paradoxical split S_2 may occur if ventricular focus leads to late depolarization of left ventricle (ECG form of LBBB). ■ Gallop rhythm frequent. ■ Murmurs and other abnormal heart sounds depend on underlying disease.

ASSOCIATED FINDINGS

■ Sudden onset of weakness, palpitation; often sweating and dyspnea. ■ Blood pressure often falls; systolic pressure varies from beat to beat. ■ Sharp large *a* waves intermittently in jugular venous pulse.
■ ECG: regular QRS complexes, usually aberrant in configuration, at a rate of 100 or more with evidence of an independent, usually slower, atrial rate.
■ Separate atrial rhythm may be hard to find on standard ECG; can be clarified by esophageal lead.
■ Tachycardia may be interrupted by beats, often early, of more normal configuration — capture and fusion beats.

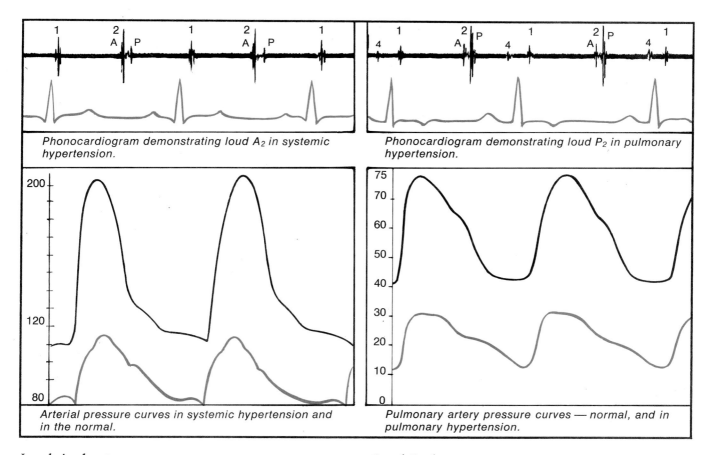

Phonocardiogram demonstrating loud A_2 in systemic hypertension.

Phonocardiogram demonstrating loud P_2 in pulmonary hypertension.

Arterial pressure curves in systemic hypertension and in the normal.

Pulmonary artery pressure curves — normal, and in pulmonary hypertension.

Loud A_2 due to
SYSTEMIC HYPERTENSION

Systemic hypertension, whether essential or secondary to renal or endocrine dysfunction, produces pressure load on the heart — signs are loud 1st sound and loud A_2. Since blood pressure is recorded routinely in adults, diagnosis should be obvious; but loud A_2 may be a clue to hypertension in children in whom blood pressure recording is rarely done.

■ Loud A_2 heard best at 2nd ICS at RSB, but transmits widely to lower LSB and apex. ■ 1st sound usually accentuated. ■ Apical presystolic gallop present without heart failure when pressure very high; both presystolic and protodiastolic gallops in heart failure.
■ Short basal ejection murmur at times.

ASSOCIATED FINDINGS

■ Positive family history common; renal disease at times. ■ Often asymptomatic; dyspnea if heart failure developing; headache and blurred vision in malignant phase. Retinal vessels show arteriolar narrowing; hemorrhages, exudate and papilledema in severe cases.
■ Blood pressure elevated, pulse vigorous. ■ Apical impulse forceful and diffuse, with palpable atrial kick at times. ■ Flank bruit in renovascular hypertension. ■ Soft pulses and low arterial pressure in legs when hypertension due to coarctation. ■ Urinalysis may show proteinuria; fixed specific gravity, casts and cells if renal disease is primary. ■ ECG: LVH, often with secondary ST-T abnormality. ■ X-ray: cardiomegaly; often dilatation of aortic arch.

Loud P_2 due to
PULMONARY HYPERTENSION

Assessment of increased intensity of the pulmonic component of the 2nd sound requires knowledge of the norms for various ages and physiques — P_2 is louder in younger and thinner people. Traditional comparison with A_2 is not reliable; a clear increase is usually solid evidence of pulmonary hypertension.

■ Loud P_2 heard best in 2nd ICS at LSB; radiates down LSB and across to RSB, but not to apex; palpable at times. ■ First sound accentuated when pulmonary hypertension is due to mitral valve disease or patent ductus. ■ Inspiratory splitting of 2nd sound may be reduced when marked pulmonary hypertension shortens right ventricular systole. ■ Presystolic gallop along LSB common unless atrial fibrillation present.
■ Short systolic ejection murmur along LSB common in all types. ■ Murmurs of underlying mitral valvular or congenital heart disease when these conditions are present.

ASSOCIATED FINDINGS

■ Breathlessness, exertional dyspnea. ■ Orthopnea in mitral valve disease. ■ Chronic cough in chronic bronchitis and emphysema. ■ Hemoptysis in mitral disease. ■ Cyanosis at times. ■ Parasternal heave.
■ Barrel chest in pulmonary emphysema. ■ Clubbing in chronic cyanosis. ■ ECG: RVH, P pulmonale or mitrale; atrial fibrillation common. ■ X-ray: enlarged pulmonary artery in all; other findings depend on primary disease.

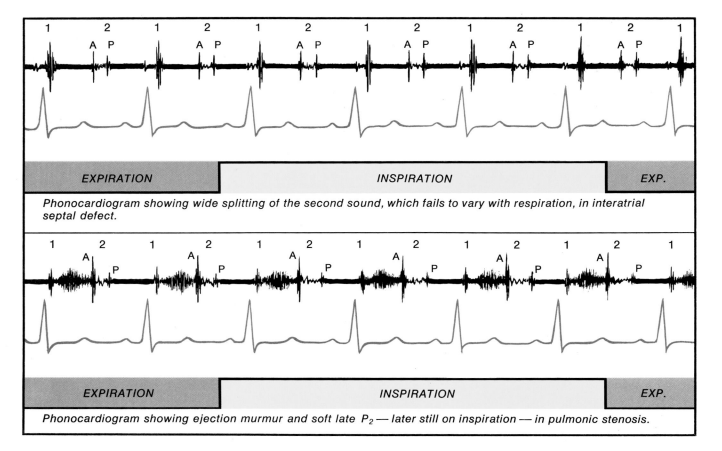

Phonocardiogram showing wide splitting of the second sound, which fails to vary with respiration, in interatrial septal defect.

Phonocardiogram showing ejection murmur and soft late P_2 — later still on inspiration — in pulmonic stenosis.

Split S_2 due to
INTERATRIAL SEPTAL DEFECT

The most valuable physical finding in interatrial septal defect is the characteristic splitting of the 2nd sound. Strenuous athletics, pregnancy, fever and anemia may simulate interatrial septal defect in the globular heart as well as ejection murmur and increased pulmonary vasculature, but they will not show fixed splitting of the 2nd sound.

- Fixed wide split of 2nd sound — does not vary between inspiration and expiration. ■ Split heard best in 3rd ICS at LSB. ■ Pulmonic component accentuated at times. ■ First sound often closely split.
- Presystolic gallop at apex. ■ Systolic ejection murmur in 2nd or 3rd ICS along LSB; no thrill.
- Mid-diastolic rumble at apex or LSB in 1/3 of cases.

ASSOCIATED FINDINGS

- Exertional dyspnea and frequent respiratory infections. ■ Left parasternal thrust. ■ Ejection sound along LSB at times. ■ ECG: diastolic overload hypertrophy of right ventricle, indistinguishable from incomplete RBBB. ■ Left axis deviation in septum primum defect. ■ P-R prolonged in 20% of cases.
- Atrial fibrillation and complete RBBB are late occurrences. ■ X-ray: cardiomegaly, enlarged arteries and relatively small aortic knob.

Split S_2 due to
PULMONIC STENOSIS

Unlike aortic stenosis, the split second sound in pulmonic stenosis can be heard readily at the bedside. It is of importance not only in supporting the diagnosis, but in assessing the severity of obstruction — the tighter the stenosis, the wider the split.

- Wide split that varies with respiration — wide on expiration, wider still on inspiration. ■ Split may be fixed with very severe lesion. ■ Pulmonic component late and soft. ■ Split heard best in the 3rd ICS at LSB.
- First sound usually normal, often followed by ejection sound. ■ Presystolic gallop along lower LSB.
- Systolic ejection murmur, often with thrill, maximal in 2nd left ICS; begins after 1st sound, may extend almost to P_2.

ASSOCIATED FINDINGS

- Exertional dyspnea or breathlessness; no orthopnea or paroxysmal nocturnal dyspnea. ■ Left parasternal lift. ■ Prominent *a* wave in neck when rhythm is sinus. ■ No moist rales, but signs of right heart failure. ■ Atrial fibrillation common; ventricular rate often rapid and difficult to control. ■ ECG: RVH and RAH, often with large late R in V_1. ■ X-ray: cardiomegaly and poststenotic dilatation of the pulmonary artery, uncommon in infundibular stenosis.

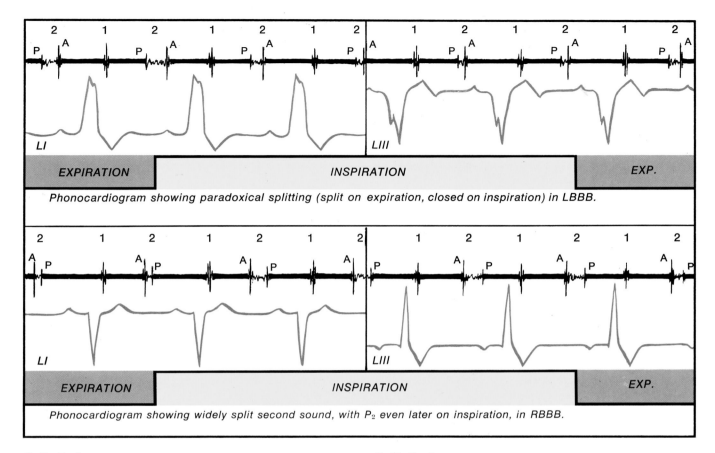

Phonocardiogram showing paradoxical splitting (split on expiration, closed on inspiration) in LBBB.

Phonocardiogram showing widely split second sound, with P_2 even later on inspiration, in RBBB.

Split S_2 due to
LEFT BUNDLE BRANCH BLOCK

Left bundle branch block is an electrocardiographic diagnosis which usually describes the functional or anatomic interruption of the left bundle branch of the A-V conducting system or all its divisions. It results in late depolarization of the left ventricle and thus splitting of the 2nd sound. Almost never seen in normals, it usually results from ASHD, but occurs in other forms of heart disease.

- Second sound paradoxically split — more split on expiration than inspiration, with the louder aortic component coming later. ■ Split usually heard best in 3rd ICS along LSB. ■ First sound often diminished with acute infarct or heart failure. ■ P_2 may be loud when heart failure present.
- Gallop rhythm often in acute infarct or heart failure.

ASSOCIATED FINDINGS

- Left bundle branch block produces no symptoms (*i.e.*, symptoms that occur are from underlying disease). ■ ECG: typically QRS is widened; 0.12 second or more; frontal QRS axis usually between +30 and −30; broad-notched R in lead I without Q or S; no Q in V_5 or V_6; notched QRS in aVL with no S wave; ST-T vectors directed opposite QRS.

Split S_2 due to
RIGHT BUNDLE BRANCH BLOCK

Perhaps the most common cause of abnormal splitting of the 2nd sound is complete right bundle branch block. While this defect can result from arteriosclerotic, congenital and other forms of heart disease, it can also be found in normal subjects with no other evidence of heart disease.

- Abnormally split on expiration; even more widely split on inspiration. ■ Late component of split is pulmonic; intensity depends on underlying disease.
- First sound sometimes split. ■ Murmurs and gallop depending on underlying disease.

ASSOCIATED FINDINGS

- No symptoms attributable to the right bundle branch block itself; no physical signs other than the typical split sound. ■ ECG shows QRS widened to 0.12 second or more with broad S wave in I, aVL, V_5, V_6 and broad R in aVR, and RSR^1 in V_1. ■ Right bundle branch block, and thus splitting, may be intermittent. ■ When ECG suggests RBBB but the splitting of the second sound is fixed, interatrial septal defect should be suspected.

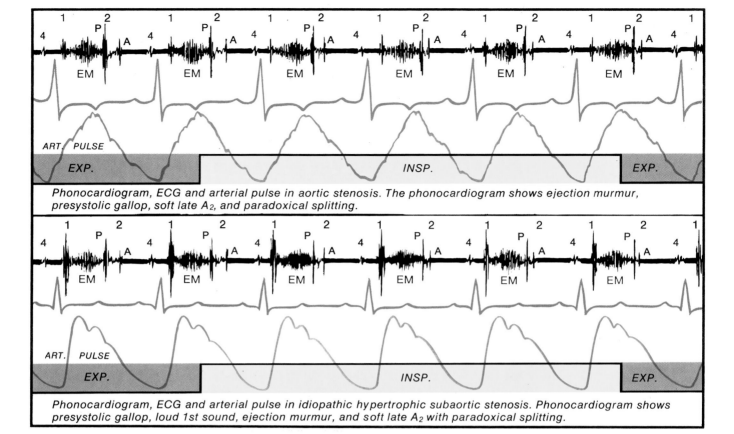

Phonocardiogram, ECG and arterial pulse in aortic stenosis. The phonocardiogram shows ejection murmur, presystolic gallop, soft late A_2, and paradoxical splitting.

Phonocardiogram, ECG and arterial pulse in idiopathic hypertrophic subaortic stenosis. Phonocardiogram shows presystolic gallop, loud 1st sound, ejection murmur, and soft late A_2 with paradoxical splitting.

Split S_2 due to AORTIC STENOSIS

Splitting of the 2nd sound is a difficult-to-diagnose physical sign in aortic stenosis, since the harsh murmur often obscures P_2 and the late A_2 is soft. When identifiable, it is valuable because it implies a hemodynamically significant degree of stenosis.

- Second sound paradoxically split, more so on expiration than inspiration. ■ Aortic component late and often abnormally soft. ■ Split usually heard best in 3rd interspace along LSB. ■ First sound increased in intensity or normal; followed by ejection sound if valve is mobile. ■ Presystolic gallop at apex frequent, even without heart failure. ■ Harsh ejection murmur at aortic area, radiating to neck and apex; often with thrill in 2nd interspace and suprasternal notch.

ASSOCIATED FINDINGS

■ Exertional dyspnea, orthopnea, paroxysmal nocturnal dyspnea; exertional angina and syncope. ■ Pulse small and slowly rising. ■ Forceful apical impulse, often with presystolic atrial kick. ■ Signs of left and right heart failure when severe. ■ ECG: evidence of LVH, usually with secondary ST-T abnormalities. ■ X-ray: cardiomegaly with rounding and enlargement of left ventricle; poststenotic dilatation of aortic root; valve calcification usually present in adults upon image-intensified fluoroscopy.

Split S_2 due to IDIOPATHIC HYPERTROPHIC SUBAORTIC STENOSIS

The finding of an unexpected split 2nd sound can be of crucial importance in suggesting idiopathic hypertrophic subaortic stenosis in a young patient with unusual dyspnea or syncope and a nondescript murmur, and is enough to warrant appropriate hemodynamic and pharmacologic studies.

■ Second sound paradoxically split at times — split greater on expiration than inspiration. ■ Aortic component late and soft. ■ Split heard best in 3rd interspace at LSB. ■ First sound normal or loud; no ejection sound. ■ Presystolic gallop with palpable atrial kick frequent; protodiastolic at times. ■ Systolic murmur, usually ejection in quality, in 3rd and 4th interspace at LSB; increased by amyl nitrite.

ASSOCIATED FINDINGS

■ Exertional dyspnea, syncope, chest pain at times. ■ Arterial pulse rises rapidly, then collapses; characteristic secondary rise prior to dicrotic notch. ■ May have prominent *a* wave in neck from associated right-sided lesion. ■ Gallop, intensity of 1st sound, murmur and bifid character of pulse all increased by isoproterenol infusion. ■ ECG: left ventricular hypertrophy, at times with Q waves suggesting myocardial infarction. ■ X-ray: cardiomegaly, but no poststenotic dilatation of aorta.

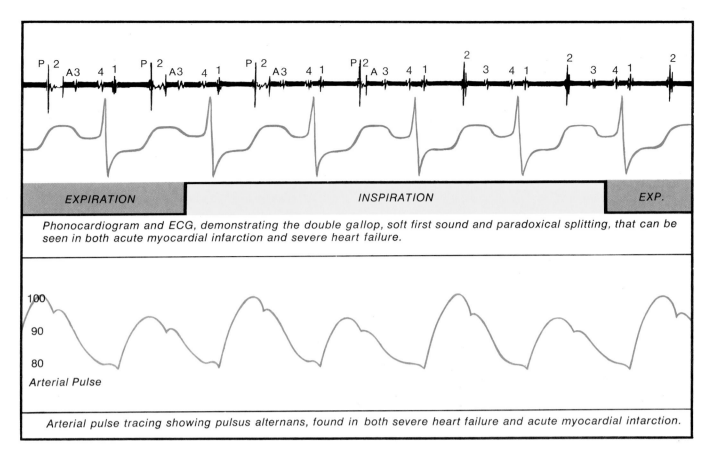

Phonocardiogram and ECG, demonstrating the double gallop, soft first sound and paradoxical splitting, that can be seen in both acute myocardial infarction and severe heart failure.

Arterial pulse tracing showing pulsus alternans, found in both severe heart failure and acute myocardial infarction.

Split S_2 due to
SEVERE HEART FAILURE

Very severe heart failure resulting in depression of left ventricular function, whether due to arteriosclerotic heart disease, primary myocardial disease or other causes, will have prolonged left ventricular systole because of reduced contractility, and will thus produce abnormal splitting.

▪ Second sound paradoxically split — split on expiration and closing on inspiration. ▪ A_2 late and often soft. ▪ P_2 usually accentuated. ▪ Gallop rhythm almost always present. ▪ May be presystolic or protodiastolic or both. ▪ Murmurs dependent on underlying disease.

ASSOCIATED FINDINGS

▪ Dyspnea, orthopnea, paroxysmal nocturnal dyspnea. ▪ Cyanosis of fingers and toes is common. ▪ Pulse pressure usually narrow, often with pulsus alternans. ▪ Rales and signs of pleural effusion in lungs. ▪ Distended neck veins, hepatomegaly and edema. ▪ ECG: heart rate usually rapid; other findings dependent on primary disease. ▪ X-ray: cardiomegaly and pulmonary vascular congestion.

Split S_2 due to
ACUTE MYOCARDIAL INFARCTION

Paradoxical splitting may be found in acute myocardial infarction because of the development of left bundle branch block; similar splitting can be a clue to infarction, without block, because of acute depression of left ventricular contractility.

▪ Second sound paradoxically split — split on expiration and closing on inspiration. ▪ A_2 late, normal in intensity unless softened by shock. ▪ P_2 often accentuated. ▪ First sound usually soft. ▪ Gallop rhythm frequently present, often with palpable atrial kick. ▪ Apical systolic murmur at times. ▪ Pericardial friction rub often present. ▪ Heart often slow at onset, usually rapid after first few hours.

ASSOCIATED FINDINGS

▪ Associated with or preceded by prolonged pressing substernal pain. ▪ Profuse perspiration. ▪ Dyspnea common; cyanosis at times. ▪ Pulse often soft and rapid; pulse pressure narrow. ▪ Moist pulmonary rales. ▪ Distended neck veins at times. ▪ ECG will show abnormal Q waves and acute ST-T abnormalities. ▪ X-ray: normal or pulmonary vascular congestion, with first infarct; often cardiomegaly if history of heart failure.

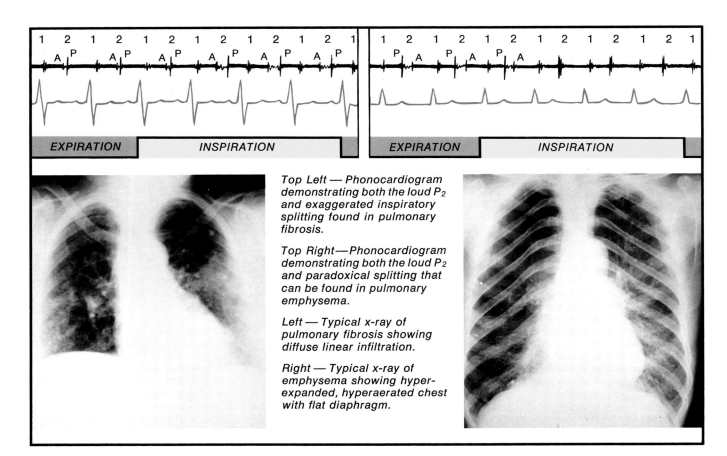

Top Left — Phonocardiogram demonstrating both the loud P_2 and exaggerated inspiratory splitting found in pulmonary fibrosis.

Top Right — Phonocardiogram demonstrating both the loud P_2 and paradoxical splitting that can be found in pulmonary emphysema.

Left — Typical x-ray of pulmonary fibrosis showing diffuse linear infiltration.

Right — Typical x-ray of emphysema showing hyperexpanded, hyperaerated chest with flat diaphragm.

Split S_2 due to PULMONARY FIBROSIS

Diffuse pulmonary fibrosis, by stiffening the lungs, increases the work of breathing and the negative intrathoracic pressure upon inspiration. This in turn increases venous return and right ventricular output on inspiration and produces the characteristic splitting of the second sound.

- Slight, normal splitting of second sound on expiration; abnormally wide split on inspiration.
- Late component is pulmonic; accentuated with pulmonary hypertension. ■ May show presystolic gallop along lower LSB with pulmonary hypertension.

ASSOCIATED FINDINGS

- Often a history of exposure to industrial dust, or of healed advanced tuberculosis or sarcoid.
- Progressive exertional, then resting dyspnea.
- Chronic cough, often nonproductive. ■ Cyanosis late, then persistent. ■ Breath sounds loud, often harsh. ■ Few fine rales at bases bilaterally.
- Parasternal lift at times. ■ Lung function will show relatively well-preserved maximum breathing capacity, markedly reduced vital capacity, often gas exchange pattern of "alveolo-capillary block."
- ECG: may show evidence of right ventricular and right atrial hypertrophy. ■ X-ray: diffuse fine infiltrate through both lungs; cardiomegaly at times.

Split S_2 due to PULMONARY EMPHYSEMA

Clear-cut splitting of the second sound is uncommon in emphysema but does occur at times; surprisingly, it is a paradoxical split — possibly due to the flattened diaphragm reducing venous inflow on inspiration in these cases.

- Split second sound is paradoxical — wide on expiration and close on inspiration. ■ Pulmonic component often accentuated. ■ Presystolic gallop at left sternal border when rhythm is sinus.
- Parasternal lift. ■ Ejection systolic murmur along left sternal border at times.

ASSOCIATED FINDINGS

- Exertional, later resting, dyspnea. ■ Cough often productive of purulent sputum. ■ Cyanosis, often worsened by intercurrent infection. ■ Prominent a wave in jugular venous pulse. ■ Barrel deformity of chest common. ■ Breath sounds diminished, with prolonged expiration, scattered wheezes and rhonchi.
- ECG: often P pulmonale; right axis deviation common; tall late R in V_1 uncommon. ■ X-ray: hyperexpanded chest; enlarged pulmonary arteries and cardiomegaly at times.

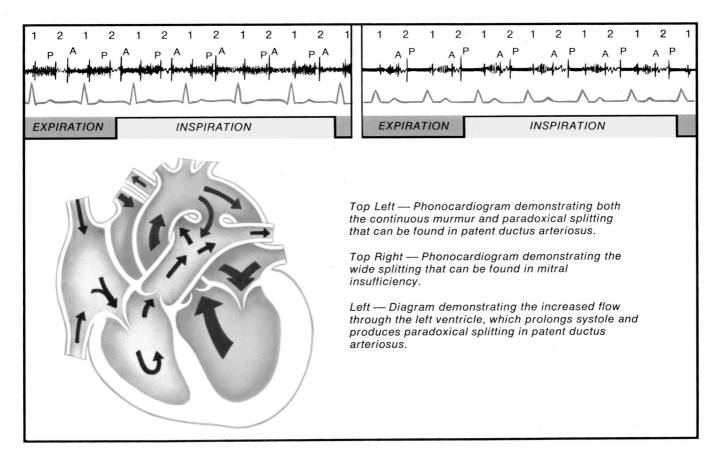

Top Left — Phonocardiogram demonstrating both the continuous murmur and paradoxical splitting that can be found in patent ductus arteriosus.

Top Right — Phonocardiogram demonstrating the wide splitting that can be found in mitral insufficiency.

Left — Diagram demonstrating the increased flow through the left ventricle, which prolongs systole and produces paradoxical splitting in patent ductus arteriosus.

Split S_2 due to PATENT DUCTUS ARTERIOSUS

Splitting of the 2nd sound occurs in patent ductus arteriosus when there is a large left to right shunt so that the left ventricle is pumping considerably more blood than the right. In such large shunts the diagnosis should be readily apparent.

- Second sound paradoxically split — more split on expiration than inspiration. ■ Aortic component is late, often loud. ■ Split usually best heard in 3rd interspace along LSB. ■ Pulmonic component loud with pulmonary hypertension. ■ First sound usually increased in intensity. ■ Protodiastolic gallop may be present at apex. ■ Continuous machinery murmur, maximal in 2nd ICS at LSB, radiating widely and often with thrill. ■ Ejection or holosystolic murmur along lower LSB. ■ Mid-diastolic rumble of functional mitral stenosis is often heard at apex.

ASSOCIATED FINDINGS

- Growth and development usually normal.
■ Recurrent pulmonary infections. ■ Dyspnea, even left heart failure. ■ Pulse bounding with pulse pressure more than 45 mm Hg. ■ ECG: LVH, often with deep narrow Q and tall peaked T in V_5 and V_6.
■ Biventricular hypertrophy with pulmonary hypertension. ■ Slight to moderate cardiomegaly; dilated aortic knob and large atria. ■ Increased pulmonary vasculature.

Split S_2 due to MITRAL INSUFFICIENCY

Prominent splitting of the second sound is not common in mitral insufficiency. When it occurs it is due to marked shortening of left ventricular systole — in turn resulting in regurgitation of a large amount of blood to the left atrium.

- Split is wide in expiration and becomes even wider on inspiration — fixed at times. ■ Pulmonic component of second sound is later — often accentuated. ■ First sound loud or normal — often obscured by murmur. ■ Harsh apical holosystolic murmur, often with thrill. ■ Apical protodiastolic gallop rhythm common. ■ Gallop may be followed by short apical mid-diastolic rumbling murmur.

ASSOCIATED FINDINGS

- Exertional dyspnea, orthopnea, paroxysmal nocturnal dyspnea. ■ Signs of left and right heart failure at times. ■ Pulse often brisk, though pulse pressure not widened. ■ Vigorous, diffuse apical impulse. ■ ECG: LVH common, associated RVH at times; P mitrale or atrial fibrillation. ■ X-ray: cardiomegaly, enlarged left atrium — increased pulmonary vasculature.

DIFFERENTIAL DIAGNOSIS OF ABNORMAL HEART SOUNDS

	Intensity of first sound	Intensity of components of second sound	Splitting of second sound	Associated Gallop
Loud S_1 due to **INCREASED CONTRACTILITY**	Accentuated	A_2 loud when systolic arterial pressure up	Normal	Presystolic gallop at apex often present; protodiastolic in aortic insufficiency and mitral insufficiency
Loud S_1 due to **MITRAL STENOSIS**	Accentuated	P_2 often accentuated	Normal or less than normal splitting	Presystolic xiphoid gallop at times
Soft S_1 due to **DECREASED CONTRACTILITY**	Soft	P_2 loud when heart failure is marked	Paradoxical when marked	Both presystolic and protodiastolic apical gallops often present
Varying S_1 due to **ATRIAL FIBRILLATION**	Varying — louder with shorter R-R intervals	Normal, unless other factors are operative	Usually normal	None unless other factors present
Varying S_1 due to **COMPLETE HEART BLOCK**	Varying even though rate is regular	A_2 normal or loud depending on systolic pressure	Variable, may be paradoxical if pacemaker has LBBB form	Independent atrial sounds present at times
Varying S_1 due to **VENTRICULAR TACHYCARDIA**	Varying even though rate is regular	A_2 may be soft with associated hypotension	May be paradoxical if tachycardia has LBBB form	Protodiastolic at times
Loud A_2 due to **SYSTEMIC HYPERTENSION**	Accentuated	A_2 accentuated	Normal	Presystolic gallop at apex — often
Loud P_2 due to **PULMONARY HYPERTENSION**	Dependent on cause of pulmonary hypertension; loud in mitral stenosis	P_2 accentuated	Normal splitting usually absent (sounds are very close)	Presystolic xiphoid gallop — often
Split S_2 due to **INTERATRIAL SEPTAL DEFECT**	Usually normal	Usually normal; P_2 loud at times	Fixed split	Presystolic at apex at times
Split S_2 due to **PULMONIC STENOSIS**	Usually normal	A_2 normal; P_2 usually soft	Late P_2, even later with inspiration	Presystolic at xiphoid often present

Associated Murmurs	Associated Signs and Symptoms	Key Laboratory Data
Typical murmur of aortic stenosis, aortic insufficiency or mitral insufficiency, when present	Pulse vigorous with steep upstroke in most; steep collapse in aortic insufficiency; vigorous apical impulse	ECG: high voltage QRS in all; definite LVH in some; x-ray: cardiomegaly in all if severe; left atrial enlargement in mitral insufficiency
Decrescendo-crescendo rumble at apex	Progressive exertional dyspnea and orthopnea; nonproductive cough, hemoptysis; signs of pulmonary edema and right failure at times	ECG: P mitrale or atrial fibrillation; RVH at times; x-ray: enlarged left atrium, pulmonary arteries
Usually none	Often with chest pain, dyspnea or sweating of acute infarct; long standing valvular disease or cardiomyopathy may be present; pulse pressure narrow and pulsus alternans at times	ECG: often recent infarct and/or prolonged P-R; x-ray: cardiomegaly frequently present
None unless other factors present	Often breathlessness and palpitation; amplitude of pulse and blood pressure vary beat to beat; other signs depend on underlying disease	ECG: totally irregular QRS with continuous atrial activity (f waves), variable in rate and form — seen best in V_1
Usually none	Intermittent syncope common; pulse pressure wide; intermittent cannon a waves in neck	ECG: regular slow QRS, rate of which is unrelated to atrial rate; QRS may be normal or aberrant
Usually none	Sudden onset of weakness, palpitation, often sweating and dyspnea; blood pressure falls; systolic pressure varies beat to beat; intermittent cannon a waves	ECG: regular QRS complexes, usually aberrant in form, rate over 100; independent from slower atrial rate
Usually none	Often asymptomatic; dyspnea with heart failure; headache and blurred vision and advanced retinopathy in malignant; blood pressure elevated, pulse vigorous	Urine — may show proteinuria, fixed specific gravity, casts, cells; ECG: LVH; x-ray: cardiomegaly, often dilatation of aortic arch
Murmurs of mitral valve or congenital heart disease, when present; ejection murmur at LSB in all	Breathlessness, exertional dyspnea; often cyanosis; parasternal heave; signs and symptoms of underlying heart or lung disease	ECG: RVH; P pulmonale or mitrale; atrial fibrillation common; x-ray: enlarged pulmonary arteries
Systolic ejection murmur along left sternal border	Exertional dyspnea and frequent respiratory infections; left parasternal thrust	ECG: simulates incomplete RBBB; left axis deviation in ostium primum; P-R prolonged in 20%; x-ray: big pulmonary arteries and cardiomegaly
Systolic ejection murmur at 2nd and 3rd ICS	Exertional dyspnea or breathlessness; prominent a waves in neck; no moist rales; often signs of right heart failure; atrial fibrillation common	ECG: RVH and RAH; often with large late R in V_1; x-ray: cardiomegaly and poststenotic dilatation of pulmonary artery

(Continued on next page.)

DIFFERENTIAL DIAGNOSIS OF ABNORMAL HEART SOUNDS (continued)

	Intensity of first sound	Intensity of components of second sound	Splitting of second sound	Associated Gallop
Split S_2 due to **LEFT BUNDLE BRANCH BLOCK**	Normal unless another cause is operative	Normal unless other factors present	Paradoxical splitting	None unless other factors present
Split S_2 due to **RIGHT BUNDLE BRANCH BLOCK**	Normal unless another cause is operative	Normal unless other factors present	Split on expiration, more so on inspiration	None unless other factors present
Split S_2 due to **AORTIC STENOSIS**	Often accentuated	A_2 usually soft	Paradoxical splitting in severe cases	Presystolic gallop at apex often present
Split S_2 due to **IDIOPATHIC HYPERTROPHIC SUBAORTIC STENOSIS**	Usually accentuated	A_2 normal or decreased	Paradoxical at times	Presystolic at times; protodiastolic gallop at apex
Split S_2 due to **SEVERE HEART FAILURE**	Usually soft	P_2 often accentuated	Paradoxical when severe	Protodiastolic and presystolic gallops common
Split S_2 due to **ACUTE MYOCARDIAL INFARCTION**	Usually soft	A_2 often soft; P_2 loud when heart failure present	Paradoxical when severe	Protodiastolic and presystolic gallops common
Split S_2 due to **PULMONARY FIBROSIS**	Usually normal	P_2 often accentuated	Normal on expiration; marked on inspiration	Presystolic at xiphoid at times
Split S_2 due to **PULMONARY EMPHYSEMA**	Usually normal or soft because of interposed lung	A_2 often soft; P_2 loud at times with pulmonary hypertension	Paradoxical at times	Presystolic at xiphoid at times
Split S_2 due to **PATENT DUCTUS ARTERIOSUS**	Usually accentuated	A_2 usually loud, P_2 loud when PA pressure is high	Paradoxical splitting in severe cases	Presystolic gallop at apex often present
Split S_2 due to **MITRAL INSUFFICIENCY**	Often loud but obscured by murmur — soft when CHF marked	P_2 often accentuated	Increased, when left ventricular systole is short	Protodiastolic at apex often present

Associated Murmurs	Associated Signs and Symptoms	Key Laboratory Data
Usually none	Signs and symptoms of underlying disease	ECG: QRS 0.12 second or more wide; broad notched R in I without Q or S; no Q in V_5 or V_6, notched QRS in aVL with no S wave
None unless other factors present	No other signs or symptoms unless due to primary disease	ECG: QRS 0.12 second or more wide with broad S in I, aVL, V_5, V_6 and broad R in aVR and RSR in V_1
Ejection systolic murmur at base of heart	Exertional dyspnea, orthopnea, PND; exertional angina and syncope; pulse small and slowly rising; forceful apical impulse, often with presystolic atrial kick	ECG: LVH; x-ray: cardiomegaly with rounding of LV; poststenotic dilatation of aortic root; calcification often seen
Ejection or nondescript murmur along lower left sternal border	Exertional dyspnea, syncope, chest pain at times; arterial pulse rises rapidly, then collapses, then secondary rise; may have prominent jugular *a* wave from associated right-sided lesion.	Gallop, intensity of 1st sound, murmur and bifid character of pulse — all increased by isoproterenol infusion; ECG: LVH; x-ray: cardiomegaly, no poststenotic dilatation
Usually none	Dyspnea, orthopnea, PND; acrocyanosis common; pulse pressure narrow, often with pulsus alternans; pulmonary rales and effusion; signs of right heart failure	ECG: depends on primary disease; x-ray: cardiomegaly and pulmonary vascular congestion
Usually none	Prolonged substernal pain, diaphoresis; dyspnea common, cyanosis at times; signs of left, often right heart failure	ECG: abnormal Q waves and acute ST-T abnormalities; x-ray: normal or vascular congestion
Usually none or short ejection; systolic murmur at left sternal border	History of industrial exposure, T.b. or sarcoid common; progressive exertional, then resting dyspnea; chronic cough; late and persistent cyanosis; breath sounds loud and harsh; few fine rales at bases	Well-preserved MBC, low vital capacity; ECG: RVH and RAH; x-ray: diffuse fine infiltrate in both lungs
Usually none	Exertional, later resting dyspnea; productive cough, purulent sputum; intermittent cyanosis; barrel chest; breath sounds diminished with prolonged expiration, wheezes and rhonchi	ECG: often P pulmonale, right axis deviation, at times; tall RV_1 uncommon; x-ray: hyperexpanded chest; big pulmonary arteries
Continuous machinery murmur at upper right sternal border	Recurrent pulmonary infection, dyspnea; left heart failure at times; pulse bounding with wide pulse pressure	ECG: often with deep narrow Q and tall T in V_5 and V_6; biventricular hypertrophy with pulmonary hypertension; x-ray: cardiomegaly with large pulmonary arteries and aorta
Holosystolic at apex; at times short mid-diastolic rumble	Exertional dyspnea, orthopnea, PND; signs of left and right heart failure; pulse often brisk; vigorous apical impulse	ECG: LVH; associated RVH at times; P mitrale or atrial fibrillation; x-ray: left ventricular and atrial enlargement

JOSEPH S. CHIARAMONTE, M.D.
649 MONTAUK HWY.
BAY SHORE, N.Y. 11706